D1087915

SCIENCE IN A
RENAISSANCE SOCIETY

History

Editor

PROFESSOR JOEL HURSTFIELD
D.LIT

Astor Professor of English History
in the University of London

SCIENCE IN A
RENAISSANCE SOCIETY

W. P. D. Wightman

Formerly Reader in the History and Philosophy of Science
University of Aberdeen

HUTCHINSON UNIVERSITY LIBRARY
LONDON

HUTCHINSON & CO (*Publishers*) LTD
3 Fitzroy Square, London W1

London Melbourne Sydney Auckland
Wellington Johannesburg Cape Town
and agencies throughout the world

First published 1972

*This book has been set in Times type, printed in Great Britain
on smooth wove paper by Anchor Press, and
bound by Wm. Brendon, both of Tiptree, Essex*

ISBN 0 09 111650 3 (cased)
0 09 111651 1 (paper)

TO THE MEMORY OF M.W.

A thing of beauty is a joy for ever:
Its loveliness increases; it will never
Pass into nothingness . . .

CONTENTS

PREFACE

The writing of this book was undertaken at the request of the Editor of the series, Professor Joel Hurstfield. Its title and scope were due to him; without his encouragement and constructive criticism it could not have achieved whatever merit it may be found to possess. I alone am responsible for the form it has assumed.

Although conceived as a historical study of the interrelations of science and society during the period *c*.1450–*c*.1620 it must not be regarded as even the outline of a history of science in that period: a few great names are entirely omitted, and the emphasis on those that are included is often not what it would be in what has been aptly called an 'internal' history of science.

The exposition presents grave problems that are certainly not wholly solved. In a period of rapid change it is desirable to maintain chronological perspective; the unqualified achievement of this aim would involve either holding in the mind simultaneously a number of disparate ideas, or the 'fragmentation' of science into the separate 'subjects' whose emergence was necessitated by the vast growth of knowledge in the subsequent 'Century of Genius'. A similar dilemma is presented by the existence of different cultures politically demarcated by fluid and sometimes arbitrary boundaries. Yet these problems seem to be of the very essence of 'science in a Renaissance society'. To mitigate the strain on the reader's attention some repetition has been consciously resorted to. An equally daunting but somewhat different problem has been that of selection having regard to the limitations of scale. This presents one of the most insidious temptations that can assail the historian of ideas, namely to embrace some general explanatory category—a change from 'Aristotelianism' to 'Platonism'; rejection of the authority of 'the Church'; or what-

ever—and then look for the evidence. While paying, I hope, a
proper respect to each of these well-publicised 'causes' of the
'transvaluation of all values' conveniently referred to as the 'Renaissance' I have with perhaps greater assiduity sought for counter-
evidence. As a consequence I have found no reason for embracing
any one of them at the expense of the others. In the kingdom of the
history of ideas there are many many mansions. We should, as I
have tried to do, inspect them all under the guidance of scholars
enamoured of their several virtues before deciding which—if any—
to live in.

Nowhere, I believe, has the general course of the argument been
made to depend on the precise understanding of a technical term. In
a few cases, however (e.g. the assessment of the aim and achievement
of Copernicus), the greater precision made possible by relatively
recent scholarship can be attained only in terms that may not be
familiar to the general reader. Such a reader may without serious
misunderstanding pass over these minor technical intrusions; but
here and elsewhere he may find it worth while to turn to the Glossary
(pp. 15–16). The 'chronological-national grid' preceding the Glossary
may help the reader to form a more 'solid' picture of the entrances
and exits of the individual actors in the drama, and the scene against
which they played. Finally, the Bibliographical Note (pp. 175–83)
should be regarded not only as a 'suggestion for further reading' but
also as a guide to the sources on which the necessarily rather condensed text has been constructed.

The foundation of this work was laid in the Library of the University of Aberdeen; the raising of most of the fabric was made possible
by two years' access to the riches of the Bodleian Library at Oxford.
The unfailing cooperation of the staff of each of these institutions
has far exceeded mere technical efficiency. I have tried in the Bibliographical Note to indicate those scholars whose works have been
a constant source of inspiration and guidance to me. My colleague
at Aberdeen, Professor W. S. Watt, Fr John Russell, s.j., and my
friend, Warren Derry, have in various places made good the inadequacy of my Latinity; and Dr Robb-Smith provided a firmer
foundation for my assessment of Renaissance medical learning at
Oxford.

The dedication to the memory of my wife reflects my indebtedness
to her knowledge of the history of architecture; but more especially
to her patience and fortitude in the course of innumerable tramps
in search of the less frequented 'memorials' of Northern Italy.
Without this collaboration I could never have begun to grasp the
complex grandeur of that marvellous age.

                                                      W.P.D.W.

CHRONOLOGICAL CHART

	ITALY	'LOW COUNTRIES'	'GERMANY'	FRANCE	ENGLAND	'IBERIA'	SWITZERLAND	OTHERS
1600	Galileo [Harvey]			Descartes	Harvey Fludd Bacon			
1575	Massaria Fabrizio Scaliger, J. J. [Clavius] Della Porta	Stevin De l'Ecluse De l'Obel Dodoëns	Clavius Ercker		Gilbert Harriot Gresham, d. Norman Dee			Tycho Brahe
1550		Mercator	Wilhelm of Hesse	Viète Paré Pelletier	Digges, T.			
1525	[Coiter] Colombo [Vesalius] Biringuccio Cardano Ghini Massa Tartaglia Da Monte	Coiter Vesalius Gemma, R.	Reinhold Rheticus [Paracelsus] Melanchthon Agricola	Rondelet Belon Ramus Fernel *Collège* *Royale* Rabelais Sylvius	Caius Edwards Recorde Turner Digges, L.	Servetus	Fuchs Bock Brunfels Paracelsus [Erasmus]	Copernicus
1500	Fracastoro Sammichele Benivieni Machiavelli Pomponazzi Manardi		Luther Agrippa		John Cabot Linacre	Magellan [Vespucci]		

		Erasmus	Dürer	Le Fèvre Chuquet	[Erasmus]	Vasco da Gama Zacuto [Columbus]
1475	Leonardo da Vinci Pico della Mirandola Pacioli Ficino Columbus Francesco di Giorgio Leoniceno					
1450	Verrocchio Federigo da Montefeltro Piero della Francesca		Regio- montanus Gutenberg Peurbach		Caxton	
1425	Alberti Toscanelli [Cusanus]	Jan V. Eyck			Duke Humfrey	Prince Henry
1400	Guarino Brunelleschi		Cusanus			

The dates are to be taken only as indicators of the periods during which the men named were active. Exact dates (where known) will be found in the Index of Names.

'National' headings indicate countries of birth. Where a man's life work has had a different national setting (e.g. Columbus) his name appears also in the appropriate column and in square brackets.

GLOSSARY

Almagest: The word is the English equivalent of the Arabic corruption of the Greek μεγιστη [συνταξις], the title of the 'great [synthesis]' written by Claudius Ptolemy (A.D. *c.* 150) and which came to bear the same relation to astronomy as 'Euclid' did to geometry. Reduced to the simplest terms its assumptions were:

(1) The Earth is the 'centre' of the World.
(2) The Sun moves at a uniform rate on a circle (the excentric) whose centre is somewhat distant from the Earth.
(3) The remaining planets (except the Moon) move on circles (epicycles) whose centres move on larger circles (deferents) centred at the excentric; but the planets themselves are represented as moving at a uniform *rate* round a separate point (equant) on the side of the excentric remote from the Earth.
(4) The Moon's motion is especially 'anomalous'.

Angular velocity: the rate at which a 'point' moves over a given arc of a curved path measured with respect to an arbitrary point, e.g. the *focus* of an elliptic orbit.

Armilla(ry sphere): a 'skeleton' sphere made up of strips of metal representing major circles of the celestial sphere, e.g. equator, ecliptic.

Ascendant: see p. 36.

Azimuth: Definitions vary somewhat for different purposes, but it may be taken as meaning the *horizontal* bearing as measured by, e.g., a theodolite.

Cartography: the science and art of map-making, i.e. the projection upon a plane surface of the spherical surface of the Earth. This is a mathematical problem of high complexity; trapezoidal (Ptolemy) and cylindrical (Mercator) projections were employed in the Renaissance.

-cento e.g. trecento, quattrocento, cinquecento. It is more common in Italy to use these convenient terms ('three hundred' etc.) rather than the misleading 'fourteenth century'.

Coplanar: 'in the same plane as'.

Cosmography: lit. 'a description of the cosmos', was commonly used as a title of early printed books describing the system of concentric spheres the lowest of which was the Earth (and its oceans) up to the 'eighth' (of the fixed stars); sometimes additional higher spheres of special astronomical significance were added. Ptolemy's *Cosmographia* was also known as *Geographia*, since it was almost wholly concerned with the terrestrial sphere. 'Geography' gradually emerged as a separate 'discipline' during the course of the sixteenth century.

Cosmology: goes beyond descriptive to causal aspects of the cosmos or 'World'. *World* always means the whole cosmos *not* the Earth.

Culverin: a large cannon.

Equant: see *Almagest*.

Excentric: see *Almagest*.

Geodesy, Geodetic: the problem of determining distances and areas on the (curved) surface of the Earth.

Hermetic philosophy: see p. 152.

Intarsie: 'pictures' built up by the juxtaposition of pieces of wood of various shapes, sizes and colours, most frequently to effect a striking *trompe l'œil* representation of solid objects.

Magnetic variation (called 'declination' in modern physics, but not in navigation)*:* the slowly changing angle between the axis of, e.g., a compass needle and the geographical meridian (N–S line.) Not to be confused with *dip*, the angle assumed by a magnetised needle mounted so as to be free to move in a *vertical* plane *before* it is magnetised.

Mitral lock gate: The two components set at 90° to one another have their apposed surfaces so shaped as to make the line of junction bisect this angle; the greater the pressure of the water the more firmly are they forced together.

Morbus gallicus: the term commonly used (except in France!) during the Renaissance for syphilis or the 'great pox'.

Polaris: the Pole-star, that indicates a point in the heavens *near* the North pole of (apparent) rotation.

Portolan(o): a marine chart from which magnetic bearings between various places can be read off; coastal outlines were indicated but not landward features.

Sexagesimal base: The base of a number-system is commonly the number of 'units' employed, e.g. ten (including zero), in the decimal system. Though not employing sixty *separate* units the 'Babylonian' sexagesimal system used '60' (represented by cuneiform characters) as a base for *calculation* ($60 = 2 \times 2 \times 3 \times 5$) as well as for the measurement of angles.

Thought experiment: an operation carried out in imagination to infer what *would* happen if certain hypotheses were actually true, e.g. Newton's diagram showing a body projected at an increasing angle until it finally becomes a satellite to the Earth.

Zodiac: see p. 37.

I

THE RENAISSANCE AND SCIENCE IN ITALY

I used to marvel and at the same time to grieve that so many excellent and superior arts and sciences from our most vigorous and antique past could now seem lacking and almost wholly lost. We know from remaining works and through references to them that they were once widespread. Painters, sculptors, architects, musicians, geometricians, theoreticians, seers and similar noble and amazing intellects are very rarely found today and there are few to praise them. Thus I believed, as many said, that Nature, the mistress of things, had grown old and tired. She no longer produced either geniuses or giants, which in her more youthful and more glorious days she had produced so marvellously and abundantly.

It may come as a surprise to the reader that at the head of a work on *science* in a Renaissance society there should stand a quotation from a little book, *On Painting* (*Della Pittura*). But the author, Leon Battista Alberti, was much more than an 'artist' in the sense in which that word is now employed. Born in Genoa in 1404 into a banking family formerly established in Florence he had shown diligence to the point of exhaustion in acquiring the most comprehensive education that the Universities of Bologna and Padua provided. Thus equipped he served in the Papal Curia in Rome (where doubtless he was inspired by the architectural triumphs of Antiquity), and in the suite of Pope Eugenius IV returned to Florence in 1434. What he 'saw and heard in our city, adorned above all others', inspired the above dedication to those with whom he was most intimately acquainted among the leading artists of Florence. Both the dedication and the text that follows it reveal an observant and wide-ranging mind which both mirrored the society in which he found himself and the direction in which he was subsequently to influence its spectacular development. The justification for singling out this most noble of the 'men of the Renaissance' as a focus for our enquiry is the frequency with which his name will appear in our pages. It will repay us first to examine the dedication in detail.

No student of the Renaissance can fail to note that the very first sentence of the dedication reveals that already in 1436 (when *On Painting* was completed) the myth of the Renaissance was already accepted. Although the term 'myth' need not necessarily be identified with a 'dead product of past ages, merely surviving as an idle narrative', it has to be admitted that in this case the almost total disappearance of artistic, mathematical, and rhetorical genius postulated by Alberti was just plain nonsense. But that the 'myth' persisted, at least as a rhetorical flourish whose sincerity it is difficult to estimate, for nearly two centuries will be shown in our subsequent studies of other societies. Nor was it new; the 'ageing' of Nature has a distinct echo of the Roman farmer brooding over the land 'forespent with length of days', as portrayed by Lucretius in his poem, *On the Nature of Things*, a manuscript of which had been recently discovered. In the High Middle Ages there was an obsessive preoccupation with a vanished 'Golden Age' of sages possessed of all knowledge then being laboriously rediscovered. What was new in the Renaissance myth was the flagrant ignoring of the glories of Gothic architecture and sculpture, of *trecento* painting, and indeed of the not so very antique rhetorician, Petrarch, and Dante. This much is therefore 'mythical' in the pejorative sense. It may well have been the wider recognition of the creation of this myth by the men of the Renaissance themselves that drove historians of the recent past to adopt one of two postures: *either* that there had never been a Renaissance, *or* there had been a succession, of which the Italian of the *quattrocento* was not the most important. The wheel has now come full circle: though a much greater degree of continuity with the Middle Ages has been revealed, yet something had happened to make the world of Niccolò Machiavelli and Leonardo da Vinci a very different seeming place from that of St Thomas Aquinas. In other words, historians seem now to be in fair agreement that if the Renaissance hadn't happened it would have been, if not absolutely necessary, at least desirable, to invent it. Perhaps we can get an inkling of the nature of this mutation from Alberti's next paragraph.

Since then I have been brought back here [Florence] from the long exile in which we Alberti have grown old—into this our city adorned above all others. I have come to understand that in many men, but especially in you Filippo, and in our close friend Donato the sculptor, and in others, like Nencio, Luca, and Masaccio, there is a genius for accomplishing every praiseworthy thing. For this they should not be slighted in favour of anyone famous or of long-standing in these arts. Therefore I believe the power of acquiring wide fame in any art or science lies in our industry and diligence more than in the times or in the gifts of nature. It must be admitted that it was less difficult for the Ancients—because they had models to imitate and from which they could learn—to come to a knowledge of those supreme arts which today are most difficult for us. Our fame ought to be much greater, then, if we discover unheard-of and never-before-seen arts and sciences, without teachers or without any model whatsoever.

Note the roll-call of those in whom there was 'a genius for accomplishing every praiseworthy thing': Filippo Brunelleschi, creator of the stupendous cupola of Santa Maria del Fiore; Donatello, whose bronze 'David' is said to be the first free-standing nude human figure since Antiquity; Lorenzo Ghiberti, whose miraculous bronze doors of the baptistry were nearing completion; Luca della Robbia, perfector of the polychrome glaze applied to reliefs in terracotta; and Masaccio, whose realistic narrative frescos in the church of the *Carmine* are commonly accounted to be the most striking single break in the tradition of painting. These five pioneers—each an incomparable master in his own craft—are compared with the Ancients; but not, as might have been expected from the first paragraph, as merely successful imitators: 'Our fame ought to be much greater, then, if we discover unheard-of and never-before-seen arts and sciences, without teachers and without any model whatsoever.' Alberti is in effect claiming that it is not only a Renaissance of a past glory, but a 'naissance' of new visions and achievements.

What are the 'new arts and sciences' to which Alberti lays claim? In attempting to answer this question we shall be in a fair way towards explicating the term 'science' in a Renaissance society.

It is important to recall that the term 'science', or its equivalent in other languages, did not normally acquire its present meaning until the seventeenth century and by no means consistently until well into the nineteenth: Michael Faraday in objecting to the title 'physicist' preferred in its place not 'scientist' but 'philosopher', understanding thereby the 'natural' part of philosophy. Four centuries earlier, when Alberti was writing, knowledge of Greek was widespread in Italy (it was taught even in some 'advanced' boys' schools) and the greater part of Greek literature had been 'discovered' for the first time. This had a profound effect on the pattern of university teaching: emphasis was less on logic and 'physics' in the 'philosophy' of the trivium (grammar, rhetoric, and philosophy) than on literature, history, and moral philosophy. This fundamental basis of the Faculty of Arts, read in the original Greek and Latin texts, constituted the *studia humanitatis* (an ideal still enshrined in the Oxford School of *Literae Humaniores*) a teacher of which came to be known as a (*h*)*umanista* by analogy with the teachers already known as *artista, canonista*, etc. From this student colloquialism the later term 'humanism' probably sprang. This is still a convenient way of expressing the aim and style of early Renaissance learning, and as such it will always be used in this book; the modern application to a cult or 'philosophy' of life would have been unintelligible to the Renaissance humanists.

In the Faculty of Arts would also have been taught some at least of the quadrivium—arithmetic, geometry, music, and astronomy—despite the fact that these studies were regarded as approximating most closely to the true nature of 'science'. The distinction between 'science' and 'art' had no necessary reference to subject-matter but only to *certainty*, that is knowledge that follows *necessarily* from unquestioned premises. On this basis geometry and arithmetic (theory of numbers, to be distinguished from mere calculation called 'algorism' or 'logistic') were as near to certainty as any secular knowledge could be. Astronomy and music (theory of harmony) being the mathematical representations of celestial phenomena ($\phi\alpha\iota\nu o\mu\epsilon\nu\alpha$=appearances) could be accepted as 'sciences' at one remove; they will be considered in greater detail in later chapters. 'Physics' on the other hand was not a science but, as its name implies (Latin *physica* from Greek $\phi\upsilon\sigma\iota\varsigma$), the study of 'nature'—the realm of coming-to-be and passing-away, sharply distinguished from the changeless 'heavens' comprising everything including and beyond the sphere of the Moon. Though this radical distinction between 'nature' and the 'heavens' was probably the most serious obstacle to the emergence of modern science Aristotle's *Physics* is far from being a mere museum piece: modern physicists no longer claim 'certainty' for their knowledge.

Faculties of Science are a comparatively recent innovation. Throughout the Renaissance, and for some time after, the nearest equivalent was the Higher Faculty of Medicine, where, notably in the universities of Northern Italy, a high importance was attached to 'method' (p. 92), 'astronomy' (p. 35), and later to human anatomy and botany. The other Higher Faculty in Italy, probably not without influence on the emergence of scientific thought, was the Faculty of Laws—Canon and Civil. The Faculty of Theology which, fostered by the endowed (1253) Collège de Sorbon, came to dominate the University of Paris did not exist in the leading universities—Bologna, Padua, Pavia—in Northern Italy.

The relatively sharp distinctions outlined above were in Alberti's day beginning to wear a bit thin. In the first place the 'arts' with which he was mainly concerned—painting above all, but also sculpture and architecture—though traditionally regarded as part of the 'mechanical' art (*armatura*) and thus excluded from the 'liberal' arts of the university, are in his book given a central position. By showing that they were based on mathematics he implicitly related them to the 'liberal' arts (and sciences). Moreover, by providing a text in elegant Ciceronian Latin as well as the lively Tuscan one addressed to his colleagues, he both placed the arts within the humanist tradition and guaranteed to his little book a place in the literature of Renaissance science. He was to follow it later by a book (*On Architecture,* printed 1485) that did a similar service for architecture. The contents of *On Painting* illustrate very well the reorientation of ideas on 'arts and sciences'. Alberti if pressed would probably have regarded the discovery or invention of linear central projection (Appendix to Chapter 3, pp. 55–6) as a new 'science', Luca della Robbia's empirical discovery of polychrome glaze as a new 'art', and his own judicious remarks on the importance of understanding the reflection, differential absorption, colour-contrast, etc. of light as a part of 'physics' relevant to the art of painting. Though, as we shall find later, Alberti's influence in many fields was widespread, hardly any part of the text embodies his own original discovery. Though the Tempio Malatestiano at Rimini, the façade of Santa Maria Novella at Florence, and the greater part of Sant' Andrea at Mantua are living testimonies to his genius as an architect, not a trace of his painting survives. There is indeed little doubt that *On Painting* was based on his critical appreciation of the work of others, pre-eminently the central projection in Brunelleschi's perspective devices (p. 55). But Brunelleschi, though the more original 'scientist' and 'technologist', had neither the linguistic skill nor the broad philosophical and political background to compose such a tract as *On Painting*, to whose influence, despite the fact that it was never printed until after its work was done, the pictures of the most eminent Italian artists bear witness.

If it now be granted that a new and powerful impetus *towards* modern scientific thought had developed within the circle of artists in Florence even before Alberti gave expression to it the question remains as to whether it was unique.

Europe in the *quattrocento* differed fundamentally from the pattern of sharply demarcated nation-states (p. 112) of the early twentieth century or the complex alignment such as exists today. Broadly speaking it comprised the huge political conglomeration known as the (Holy Roman) Empire of which 'Germany' was the largest territorial bloc, the Iberian Peninsula comprising Spain and Portugal, the kingdoms of France, England, Scotland, Denmark, Sweden, Poland, Bohemia and Hungary, and the Free Cantons of Switzerland. Frontiers had none of the rigidity demanded by customs barriers. Remnants of ecclesiastical and feudal power were far from insignificant: the power of the Duchy of Burgundy was comparable with that of the surrounding Kingdom of France; the course of events in Italy—and far beyond the peninsula—could be dramatically altered by the political alignment of the Vicar of Christ. Above all, the concept of 'nationality' was in the making. On the other hand communication between widely separated cultural centres, though slow, was in many respects less obstructed than in our time. In the face of this complexity simple judgments as to the 'where and when' of scientific trends in the Renaissance can produce endless confusion. Nevertheless, Alberti's claim for the uniqueness of the Florentine impulse can be accepted as broadly true in respect of the time in which he was writing. It is also broadly true that, apart from Southern Germany and the Iberian Peninsula, only in other centres of Northern Italy was there scientific activity of even comparable importance to that of Florence. Nearly a century was to pass before the contributions of France and England became significant and before Denmark and the Low Countries were to give birth to men whose achievements form part of science as we know it today.

In the light of this assessment and in order to preserve some degree of chronological perspective the development in Southern Germany complementary to that in Italy will be dealt with in the next chapter; the restricted but decisive activity in Portugal will be considered in the chapter on the discovery of the 'New World'. The delayed but subsequently outstanding achievement in England and France having been, it is suggested, conditioned by a sociopolitical structure markedly different from that of Italy and Germany will be reviewed only after the main lines of science in a Renaissance society have been examined. One event, however, was part of both Italian and German cultural history: this was the Council of Florence held in 1439 in a last supreme effort to unite the Roman and Eastern wings of Christendom, but which had repercussions beyond the purely ecclesiastical issues involved.

The choice of Florence as the final venue of the Council was largely an accident. The Greeks might never have gone to Italy at all since they assembled first in Basel. When the deliberations there were apparently bogged down Pope Eugenius IV insisted on their transfer to Ferrara where they might have remained but for an outbreak of plague. The choice of Ferrara may well have been partly motivated by the vigorous academic life brought into being by the reigning prince Nicolò III d'Este, who installed Guarino of Verona as tutor to the heir apparent, Leonello. In Guarino's school, opened in 1436, the sons of many of the leading scholars, chancery officials, and rulers of Italy, were educated, as in Vittorino da Feltre's school in not far distant Mantua, seat of the Gonzaga family. Natural philosophy indeed (or at least the more rigorous and formal parts of it) was not regarded as a necessary accomplishment of privileged youth; nevertheless Vittorino had shown himself just as capable of rapid mastery of the contents of Euclid's *Elements* as he (and of course Guarino) had of the language in which they had originally appeared.

This emphasis on the classical languages, declamation, athletic prowess, and plain living—*mens sana in corpore sano*—in the education of the élite has been put forward as a 'proof' that the Italian Renaissance, so far from releasing the progress of science from the thraldom of the scholastic dialectic concerning the nature of motion and the like, actually retarded it. While the humanistic obsession with linguistic erudition, embodying a parade of citations from the ancient authorities, was sometimes given greater weight than independent enquiry, this was certainly not the aim of Guarino, whose motto was *nec verbum ex verbo tantisper sed sensa tantisper exprimes quasi corpus non membra circumscribas* ('Nor will you translate word for word to that extent and no more, but to the full extent of the sense, as if you would delineate its body not its individual limbs'). Even if it be allowed, as it must, that humanism favoured a backward-looking mentality, yet by insisting on an accurate and completed review of what the Ancients had taught it brought to light a fund of wisdom in relation to concrete instances. It also corrected the distorted view (even as late as Petrarch's) of the Great Men of the past as having been cast in a godlike mould above the petty concern inseparable from the market place and even the council chamber, revealing them rather as personally involved in the sort of problems that have beset men wherever questions of power and its abuse have arisen. The influence of such knowledge is outstanding in the method employed by that arch-realist, Niccolò Machiavelli, to drive home his perhaps emotionally inspired guide to perplexed rulers, *The Prince* (p. 76). The term 'humanism', of course, reflects the fact that the literary Renaissance became in large measure *Studia Humanitatis*.

Such fruits may be a far remove from 'science' in its more usual connotation; but that it was not so far removed as is commonly made out will become clearer when in a later chapter we examine the question as to how far those who had profited by the humanistic ideals of Guarino gave to Ferrara a temporary pre-eminence in Medicine. That the practice of Medicine was not then or for a long time afterwards very scientific must be readily agreed; but the problems it gave rise to were such that into those aspects of natural knowledge not covered by 'pure' astronomy it was during the same long period the only entry.

The effort to appraise the relations of 'science' and 'Renaissance' has already revealed something of the complexity of the 'society' in which these relations were set; this must now be further enquired into.

When Alberti finished writing *On Painting* he was able to report the completion by Brunelleschi of the cupola of Santa Maria del Fiore. But in relation to an undertaking believed by many of the Florentine council to be beyond the power of human genius it is appropriate to gauge the state of Florentine society in the years immediately prior to the acceptance of the task. Work in fact had started in 1420: in this year the Florentines were threatened by a new move for territorial aggrandisement by the Milanese Filippo Maria Visconti, from whose father's grasp they had, as if by a miracle, escaped probable extinction as an independent power less than twenty years earlier. Whatever may be the final decision of historians in regard to the thesis, massively maintained by Professor Hans Baron, that the stubborn resistance of the Florentines against the boundless ambition of Giangaleazzo Visconti was their finest hour, the period in which it occurred forms a valuable watershed from which to survey the most significant features within the anarchic structure of Italian society.

If, as Metternich thought, Italy was in 1849 a 'geographical expression' it was in some respects no more than this in 1400. On the other hand it was then nearer to its ancient Roman heritage and had been less mangled by the 'colonialism' of France, Spain, and Austria. Such hierarchical feudal structure as there had been had largely disappeared centuries before, the only comparable powers being the suzerainty exercised by the Empire and the Church dividing loyalties between Guelphs and Ghibellines respectively. The south was dominated by the Kingdom of Naples, the north by Milan, Florence and Venice; between them were the areas over which the Church exercised greater or less territorial sway; these included one important outlier north of the Apennines (that, of course, formed no boundary between 'north' and 'south'), namely Bologna. In addition to the 'big Four' there were scores of 'city-states', scattered in a fairly random manner, for whose allegiance or conquest the larger territorial powers continually struggled; reference has already been made to Mantua and Ferrara. Some account of the structure of all these 'communities'—a convenient word which avoids at this stage the question-begging terms 'communes', 'republics', 'despotisms' and the like—is forced upon us by the continuing debate as to whether 'democratic freedom' or 'despotic patronage' was most conducive to the progress of science in the Renaissance. Unless these terms are brought into unambiguous correspondence with actual institutions the argument is unlikely to be resolved.

Not unexpectedly, evidence of communal spirit preceded any kind of constitutional instrument. As early as 1081 the Emperor Henry IV recognised in the maritime power of Pisa a governing body whose officers called themselves 'consuls'; the title *commune colloquium civitatis* ('common parliament of the state' in the literal sense of a common 'talking-ground') was the justification of the later application of the term 'commune' to similar self-conscious formations that rapidly multiplied during the thirteenth century. The frequent attribution of 'democratic' republican aims to these bodies is far more problematic: they were indeed not 'despotisms', but neither were their 'senates' at all widely representative. In the first place all power was concentrated in the 'city'; to the inhabitants of the surrounding country they provided a market and in turn must have recognised their vital dependence on the former for the means of existence; but in the periodic raids from the neighbouring communes the country folk would be the first to suffer. Nor did more than a comparatively small minority of the townsfolk have any right to express any opinion; issues were likely to be decided to the advantage of the leading merchants and more powerful guilds. Nevertheless, one by one the communes fell under the autocratic sway of some ambitious man who had established his eminence not by the promotion of industry and commerce but by the rape of neighbouring communes; thus as early as about 1330 the Scaligeri of Verona had pushed their conquests as far as Lucca, a key town on the road from Florence to Pisa. Well before the end of the *trecento* there commenced the most ambitious enterprise of all which, had it succeeded, would have brought the whole of Northern Italy including Bologna (though this was normally under the suzerainty of Rome) under one centralised power, that of Giangaleazzo Visconti. The enterprise was indeed a model of inhuman dedication to political ends, carried through with superhuman efficiency, based on nice political calculation supported by statistically controlled logistic. Here if we are not to be too nice in the use of the term 'science' we may see an early example of scientific warfare as the 'continuation of diplomacy by other means', the explicit statement of whose theory had to wait for Clausewitz in the nineteenth century.

With all this ruthless lust for power went a concern for the arts and sciences that found its highest expression in Giangaleazzo's raising of the rather obscure *studium generale* at Pavia to rank with the most distinguished universities of the peninsula—Bologna and Padua. But, as Garrett Mattingly pointed out, this gesture may have been motivated less by a concern for the liberal arts as such than by *raison d'État;* for the main emphasis was on the Law School. We have already noted (p. 22) similar acts of patronage by the ruling princes of Ferrara and Mantua. On the other hand in Florence— paragon of 'free' republics—the university was of short duration and gained no outstanding renown. Later (p. 46) we shall see how important for the progress of science was the patronage of despots.

In comparative isolation ever since the 'Dark Age', when they sought refuge from the incursions of barbarous hordes on the mainland, the inhabitants of Venice evolved a system of government that does not fit neatly into either 'commune' or 'despotism'. Though the doges had near-despotic power, any kind of *dynastic* absolutism, such as characterised most notably the d'Este of Ferrara, was excluded by a limit to the time during which the office might be held; and furthermore the domination of family cliques, such as periodically threw many of the mainland cities into bloody chaos, was prevented by the use of the lot and an elaborate scheme of checks and balances which constitute an interesting case-history in political science. Though Venice was not a notable centre of scientific activity it must not be forgotten that, facing eastward as she does, she was strongly influenced by Byzantine culture; and it was there in 1403 that Guarino (p. 22) met Manuel Chrysoloras in the suite of the Eastern Emperor and accompanied him to Constantinople. And it was to Venice that, after a brief spell of teaching in Florence, he returned to set up his own school where he taught Greek to the majority of those Italians who became the leading scholars of their age.

Though her control of the trade routes from the East was being threatened by the advance of the Ottoman Turks she had at this time displaced the Carrare as 'Lords' of Padua, and, apart from the years of the near-domination of Giangaleazzo, exercised an overall rule of the cities of Verona and Vicenza. From Verona the road runs due north through Trento over the Brenner Pass to Innsbruck, and thus to the great centres of German burger patrician culture, Augsburg (the headquarters of the Fugger family), Ulm and Nürnberg. From Venice itself, a road runs north through Treviso and Udine over the Carnic Alps to Villach and Klagenfurth in Carinthia, thence to Vienna and Cracow. In the next chapter we shall attempt to assess the relationship between those regions and Italy of the *quattrocento*.

2

SCIENCE AND THE NORTHERN
RENAISSANCE

The roads from Venice and Verona to the countries of Central
Europe were two-way traffic routes: in respect of the linguistic
Renaissance—the cult of Latin and Greek classics—the traffic was
almost wholly northward; where science and mathematics were
concerned the southward drift of men may obscure the consequent
movement of ideas: many who went to Ferrara, Bologna, and Padua
to learn stayed to teach. Of none could this judgment be more
confidently asserted than of Nicholas Kryppfs, most commonly,
though somewhat inappropriately, known as Nicholas of 'Cusa'.

Born in 1401 at Cues, a small township linked by a bridge across
the Mosel with the better known Berncastel, Nicholas received his
early schooling at Deventer, where the Brethren of the Common
Life had aimed at liberalising the more arid instruction of the
cathedral schools. He matriculated at Heidelberg in 1416 and subsequently made his way to Padua, where in 1423 he received the
Doctorate in Laws. There being no Faculty of Theology at Padua,
and its dominant philosophy being somewhat suspect, Nicholas,
who was to attain his ecclesiastical office as bishop and cardinal
legate, completed his studies at the University of Cologne. With
Nicholas's ecclesiastical career we are not directly concerned; but
two events widely separated in space and time were decisive for his
subsequently appearing as one of the few key figures in the emergence
of new scientific attitudes in the *quattrocento*: his close association
with the Greek party on the voyage from Constantinople and
subsequently at Florence; and, as Bishop of Brixen (now Bressanone
in Italy), his journeys as papal legate throughout the western
territories of the Empire.

It was during the voyage from Constantinople that Nicholas

wrote a great part of his *De Docta Ignorantia* (*Learned Ignorance*) the work by which he is most widely known. Its aim was wholly theological—to clarify the relation of Man to God and to the rest of God's creation. But in the prosecution of this aim he made use of two principles closely bound up with the nature of scientific knowledge, namely, that knowledge of nothing is certain or complete, and the best way to approximate the unknowable truth is by way of that human discipline which most closely approximates certainty, mathematics. Knowledge is to be approached not so much by new speculation as by the affirmation of the limits of human understanding; ignorance indeed, but *learned* ignorance. Basic to this learned ignorance is the recognition of the relativity of all finite experience in relation to the infinite creation; 'infinity' is here to be taken literally, namely, as 'unbounded': to set bounds to the creation is to derogate from the omnipotence of the Creator. This insight—and the dilemmas it gave rise to—had been made explicit by Abelard in the twelfth century; but no one before Nicholas had both squarely faced and accepted the cosmological implication that unless the world is a *finite* sphere it cannot be apprehended as a sphere at all. For in the infinite sphere centre and circumference are indistinguishable, in that any point may with equal justice be regarded as centre. If this be so the Aristotelian universe of shells (or 'orbs') concentric with the Earth becomes untenable, as does the concept of a stationary Earth about which all the celestial bodies are carried round by their orbs.

It is important to note that the belief (still to be found) that Nicholas in effect, if not actually in so many words, 'anticipated' Copernicus in putting the Earth in motion *round the Sun* betrays a complete misunderstanding—and diminution—of his teaching, the central feature of which was the indifference of *all* points: if any move, as many evidently do, then so do they all. It is only the failure of 'ignorance' to become 'learned' that permits us to insist on the absolute rest of any body. If Nicholas 'anticipated' anyone, it was not Copernicus but Henri Poincaré and Einstein. But to make such a claim is equally to misconceive Nicholas's general standpoint. Though he collected the best instruments then obtainable (they may still be seen in his study preserved as a museum in Cues) he was no systematic observer of the skies; by pressing his geometrical analogies too far he revealed his mathematical inferiority to some of his contemporaries. The apparently 'modern' outcome of his speculation followed from his concern for theological truth: there was no single strand in his argument—even that of the 'coincidence of opposites' in the 'maximum'—that had not been envisaged during the High Middle Ages or even in Classical Antiquity. His claim to being one of the great pioneers—perhaps even *the* pioneer—of the new ways of thinking about Nature characteristic of the Renaissance lies in the application of critical imagination to existing knowledge. In this respect he may have owed more to the mystical strain characteristic of German thought at this time and to the University of Cologne with its bias towards the speculative mathematical cosmology of Plato rather than to the more matter of fact and systematic study of 'method' in the stronghold of Aristotelianism, Padua. But so far we have told only half—if perhaps the greater half—of his story.

Although it is impossible to make any direct assessment we ought not to overlook the possible influence on the emergence of science of the organisation of thought that brought about the great Law School of Bologna and later of Padua, Pavia, and Ferrara. Had not the scholar and Chancellor of Florence, Coluccio Salutati, a few years before Nicholas's birth, written a famous tract comparing the *Nobility of Laws and Medicine*? Though he plumped in the end for Laws, this was not merely the prejudiced decision of a literary humanist—he dealt knowledgeably with the great medical writers of the past including the Arabic writer Avicenna, but significantly set great store on the higher degree of *proof* attained by the logic of *propter quid* used by the lawyers as compared with the *quia* of the medical doctors (pp. 92f.)

We can hardly doubt, however, that more decisive than Nicholas's study in the Law Schools at Padua was his contact with some of the most eager mathematical thinkers of the age. Of these, one must almost certainly have been Prosdocimo de' Beldomandi, to whose teaching spanning the years spent by Nicholas at Padua generous tributes were paid. Prosdocimo left no permanent mark on the subject, but a younger man of whom Nicholas later spoke as 'having been bound from the years of our youth and adolescence' has been regarded as the leading mathematician of the age: this was Paolo dal Pozzo, usually known as Toscanelli, who in the secular pragmatic world of the *quattrocento* had as wide an influence as had Nicholas in the theological and philosophical. Here then was perhaps the first notable link to be forged between the maturing Italian Renaissance and the as yet unformed German. Some decades later they were to be closely associated with two other mathematical thinkers from Germany; meanwhile Toscanelli had come to know Alberti and had taught Brunelleschi the mathematics underlying the theory of linear perspective.

Let us now turn to look at this 'Germany' eager to share in the Italian Renaissance but having some outstanding contributions to make from its native talent. The 'German Empire' had by then lost any vestige of control over the extreme northern and southern territories, and over Switzerland and North-Western Italy its hold was only nominal. Lacking the central power of the papacy the very numerous ecclesiastical magnates made for a lack of internal cohesion even more marked than in Italy. None of the courts of the lay magnates were noted for their patronage of art or literature; when the Italian Renaissance took root it was in cities where merchant-bankers had grown to wealth and eminence: in none of these was there a university. Germany did not, however, lack seats of learning: Heidelberg and Cologne have already been mentioned as contributing to the education of Nicholas of Cues; Erfurt was probably the most progressive—Martin Luther was its most famous product. If we include the neighbouring kingdom of Bohemia, whose association with the central bloc of Germany was always very close, Prague had the most venerable foundation (1347); Vienna following in 1384 as a result of a somewhat earlier migration of German students from Paris. The nationalistic schism associated with the Hussite 'heresy' led to a growth in prestige of Vienna and the foundation of a new university at Leipzig (1409), both at the expense of Prague, whose pre-eminent reputation declined. Just within the territory of the neighbouring kingdom of Poland Cracow was re-founded in 1397, and despite its relative geographical isolation maintained high international reputation. Within the central German bloc six more universities, of which Tübingen and Ingolstadt had probably the most significance for science, were founded before the end of the fifteenth century. More notable in the Empire than in Italy were several ancient monasteries—Fulda, Reichenau, St Gallen, Reichenbach, Spanheim, Tegernsee and, less well known but for our purpose outstanding, Klosterneuburg. It was in the *armaria* (store cupboards) of such communities that Nicholas of Cues claimed that the Germans (*Alemanni*) were discovering *original* sources of ancient knowledge—'worthier of study than mere compilations written in a more elegant style'.

The significance of Klosterneuburg for Renaissance science became evident during the priorate of Georg Mustinger who is known to have sent a canon to buy books in Padua as early as 1421, and to have attended, with representatives from Vienna, the Council of Basel at the time that Nicholas of Cues was there. He collaborated with John of Gmünden, of the University of Vienna, who prepared trigonometrical tables; but probably his most important function was to have established at Klosterneuburg a reputation for mathematical studies that drew scholars from other monastic houses. Among such scholars, formerly of Reichenbach but from 1456 associated with Tegernsee, was Nicolaus Germanus—a name familiar to historians of cartography, since his dedication of the first modernised version of Ptolemy's *Cosmography* to Pope Paul II (1464–71) may be read in the edition printed at Ulm in 1482. At Tegernsee he probably met Nicholas of Cues, who was wont to regard Tegernsee as a haven of refuge from his episcopal duties, and who compiled some of his mathematical tracts at the request of its inmates. By 1466 Nicolaus Germanus had settled in Italy where he was for some time under the patronage of Borso d'Este of Ferrara. His claim to have invented the trapezoidal projection, that henceforward marked a great technical advance in map-making, seems to have been gratuitous, since examples of its earlier use at Reichenbach have since been discovered; but his application of the method in the preparation of several 'modern' maps cannot be questioned.

Just as the problem of pictorial perspective was a powerful stimulus to the cultivation of mathematical skill among the Italian 'artists', so that of cartography seems to have been the main concern of German scholars. The two problems are of course closely related: the former aims at a 'scientific' representation of space, the latter at a similar presentation of a curved surface (the Earth's) on a plane. It may seem strange that the Germans, for whom one might suppose the problem of marine navigation hardly existed, should have been first in the field with 'scientific' as distinct from descriptive mapping. Several factors may have contributed to this. The Italians (and Catalans) had in fact achieved all that was necessary for Mediterranean navigation by means of the *Portolani*—charts of compass bearings ignoring the curvature of the Earth. On the other hand the making of accurate topographical maps had depended from the time of Ptolemy—or before—on accurate observations of the Sun and stars; and Vienna had developed a strong tradition of astronomy by the time of John of Gmünden. The instrument known as the *torquetum*, which Nicholas of Cues added to his collection, had very probably been made by a clockmaker of Nürnberg in 1434. Lastly, there were economic drives which, somewhat later than Nicolaus Germanus' busiest years, became manifest among the Germans by a concern for oceanic travel with its dependence on accurate measurement of arcs of the terrestrial sphere; how far these may have been latent in the first half of the fifteenth century we have no knowledge.

When John of Gmünden died in 1442 the mathematical tradition of the University of Vienna was assured: for, though his originality has been questioned, he left to his colleagues and pupils a collection of books and apparatus. The first to take effective advantage of this, though he may have arrived too late to 'hear' John himself, was Georg (of) Peurbach: with him a link with Italian humanism was forged, and since to his high competence in Greek and mathematics was added an appreciation of the importance of systematic astrono-mical tables, there at last was a scholar capable of producing a definitive text of Ptolemy's *Almagest* such as the West had never possessed. Dying at a relatively early age he left the completion of the task to his young and even more brilliant pupil and collaborator, Johann Mueller, usually known from his birthplace, Koenigsberg (in Franconia), as 'Ioannes de Regiomonte' or simply 'Regiomontanus'. It was fortunate that at the time of the conception of this task the Greek cardinal Bessarion, who had made numerous contacts with Italian scholars (including Toscanelli) at the Council of Florence, had settled in Italy (p. 22). In the dedication to Bessarion of his *Epitome* of the *Almagest* Regiomontanus quotes the dying Peurbach as laying upon him this sacred trust in order that 'I may satisfy the desire of our best and most worthy lord, Bessarion'. The Cardinal had in fact provided Peurbach with the Greek text (salvaged from Byzantium) on which the *Epitome* was based.

Of the close-knit nature of the dialogue established between mathematically minded Italians and Germans about the middle of the fifteenth century we have the most direct evidence in a small book not published until nearly seventy years after the events it records. Some delay would have been due to the fact that the pioneer printers would hardly have squandered their slender resources on a volume appealing to only a very small number of specialists; so protracted a delay was probably due to the fact that on the death of Regiomontanus, in whose library the publisher found the manuscript, his effects passed into the hands of his young patron, Bernhard Walther, who put difficulties in the way of those wishing to profit by the great astronomer's work. The matter with which it was concerned was the unduly optimistic assumption of Nicholas of Cues that he had solved the problem of 'squaring the circle'—one that had baffled the finest minds of Greece and was later (1882) shown to be impossible of solution. It was traditionally one of three 'insolubles', and is more accurately enunciated as the problem of describing *with straight edge and compass alone* a square equal in area to a given circle; the restriction is not arbitrary but embodies at the constructive level the axioms of Euclidean geometry. There are three points of special significance for our purpose: Nicholas's supposed 'proof' was given literary expression in a dialogue (dated Brixen 1457) between himself and Toscanelli; the German 'professional' mathematician, Regiomontanus, addressed his disparaging refutation of the 'proof' to Toscanelli; and he dated it June 1464 from Venice 'when this Christian republic had been thrown into confusion (*turbata*) by its enemy, Mahomet'—this was, of course, Mohammed II, who, after storming Constantinople in 1453, had gained possession of some of the Venetian dependencies on the Dalmatian coast.

Seven years after addressing the above tracts to Toscanelli Regiomontanus was back in Germany; but this time in Nürnberg, whither, he said, he was drawn by the presence there of an active guild of mathematical (including astronomical) instrument makers. This may well have been true; Nürnberg was the nearest approach in Southern Germany to an Italian city-state with a large communal area under its jurisdiction; gun-founding and goldsmithery were already established, and before the end of the century the famous Nürnberg pocket watches were available. But a careful comparison of dates reveals the fact that the activities that made the city internationally famous came after the death of Regiomontanus in 1476: it has been plausibly urged that the social eminence of Nürnberg owed more to the fame and enterprise of the mathematician than *vice versa*. It is more likely that Regiomontanus had an additional inducement: he had to earn his living. What more suitable means than the sale of the tables necessary for astrological prediction? How much all this may have weighed with him we cannot say: what is beyond dispute is the fact that, apart from a diatribe against a late medieval work on planetary theory, the only works of Regiomontanus to be published before his death consisted of a couple of fairly extensive collections of tables, and a flood of calendars, tables of new and full Moons, and actual prognostications. But this is not all: in the annals of astrology two names stand out as having made striking innovations in the theory of prognostication—the Arabic writer Albumazer and Regiomontanus.

To the question what has astrology to do with science, the answer is that until well into the seventeenth century it had, whether for good or ill, a great deal to do with it. This relationship, more significant during the Renaissance than at any time before or since, was by no means restricted to Germany; but there is reason to believe that concern for prognostication was then more widespread and obsessive in Germany than elsewhere. If this was so it would provide another motive for the return of Regiomontanus to Germany.

In his great work on *Germany before the Reformation* Willy Andreas suggests that the seed that developed into so rank a growth was in fact brought thither by scholars returning from Italy. Those who had studied at Padua could hardly have failed to see the imposing array of astrological symbols surrounding the huge sala of the Palazzo della Ragione. These may well have been inspired by the works of Pietro d'Abano to whose teaching the rise of the Medical School may be fairly attributed. Henceforward in the practice of medicine a rigorous analysis of method (see Chapter 7) was found to be compatible with a systematic study of astrological 'influences'. As late as about 1500 Baldassare Peruzzi thought it worthwhile to decorate the Farnesina in Rome with an elaborate set of frescos illustrating the horoscope of his patron, Agostino Chigi the banker, for whom he had built the palace. Some years before that, however, a strong current of scepticism (p.151) had begun to flow; in Germany

B

the reverse seems to have been the case. From Germany came the confident prediction of a second 'flood' (p. 38), and at the end of the century the man who far more than Copernicus effected the change to *modern* astronomy, Iohannes Kepler, played as important a part in the planning of the campaigns of Wallenstein as did the military commanders. That Kepler's scientific thought was greatly influenced by a belief in a mystical consonance between 'bodies celestial and bodies terrestrial' there is no doubt whatever: that he took very seriously the prognostications he was paid to produce is more doubtful.

Astrological activity did not, of course, die out in Italy. Paul III (to whom Copernicus dedicated his epoch-making treatise) employed Luca Gaurico to construct the horoscope of Martin Luther; his consequent assurance that Luther must have gone to hell was questioned by Girolamo Cardano (p. 139) on the ground not that such a conculsion could not be drawn from natural phenomena but that Gaurico had faked the time of Luther's birth. Neither in France nor in England do we find so intimate a connection between astrological science, natural science, and the forecasting of consequences of importance to society; but the disastrous course of events in sixteenth-century France may have been affected by the uncritical reliance on astrological portents of the Italian queen-mother, Catherine de'Medici, who from 1560 virtually ruled France for nearly thirty years.

To the general concern for judicial astrology no better clue could be found than the enormous distribution of the 'best seller' known as the *Lichtenberg Prognostications*. This was first printed in 1488 (probably at Strassburg), went through at least six editions before 1500 and eight in the following century. It was written in Latin but the later editions included versions in German and other vernaculars. It was the work of Johann Ruth (or Roth), native of Lichtenberg who lived mainly during the reign (1440–93) of the Emperor Frederick III to whom he was probably Court Astrologer. The book claimed to set out a number of celestial influences (*influxus*) and the tendency (*inclinatio*) of certain constellations, namely conjugations and eclipses, that had taken place in those years referred to, which 'portended evil or good for the world at this time and in the future and will last for further years'. Luther was no enthusiast for judicial astronomy—he found the lucubrations of his more humanistically inclined lieutenant, Peter Melanchthon, theologically somewhat embarrassing—but, realising the widespread influence of the *Lichtenberg Prognostications* he wrote a critical Foreword to the German edition of 1527 (Wittenberg). 'What are we to say, then, about Lichtenberg and others of the same sort?' he asked. 'For my part I hold his astrology (*Sternkunst*) to be correct; nevertheless the art itself is uncertain; the signs in heaven and on earth do not themselves lack certainty; but they are the work of God and the angels, and the art by which this may be apprehended is not to be found in the stars as such.' This judgment is a variant of a general attitude, adopted by many leading figures throughout the Renaissance period, which was acceptable to the Roman Church even after the onset of the Counter-Reformation.

If no clear line can be drawn between astronomy and astrology during the period with which we are concerned it seems desirable to use Luther's judgment as a text on which an elucidation of this complex relationship may be based in the light of our own concept of science.

Remembering that Luther's term translated above as 'art' was *Kunst* we first recall (p. 20) that every non-demonstrative study, however systematically undertaken, would have been regarded as an 'art'—the term 'science' being generally restricted to the *necessary* truths embodied in syllogistic logic and pure mathematics. As we have seen, and shall later find to be of the highest significance, the study of astrology was nothing if not systematic, rigorous in its inferential procedures, and demanding a knowledge of the most advanced mathematics available. It is doubtful whether any of the leading minds in the Renaissance would have rejected the 'art' of astrology as such. The paramount influence of the Sun on the whole living world was beyond question; that of the Moon, in respect of the tides, eclipses, and menstrual cycle, only less so. But these two bodies were only the most prominent of those celestial bodies that to an Earth-centred intelligence proceed in a rather complex manner. The great stellar backdrop against which the drama of the planets is played must have some significance: the distribution is anything but uniform, but this spatial relationship is the only *usable* characteristic by which 'one star differeth from another in glory'. This analysis of the intellectual origin of astrology is nowhere expressed in so many words; but is a plausible conjecture based on the so-called *Tetrabiblos*, in which Claudius Ptolemy (whose *Almagest* was the astronomical Bible of our period) set out the 'means of prediction through astronomy . . . and by means of the natural character of these aspects themselves we investigate the changes which they bring about in that which they surround'.

In the absence of any rational quantitative mechanics (established only in the seventeenth century) the art of astronomy was the nearest approach to the demonstrative science of pure mathematics—indeed we find the terms 'mathematician' and 'astronomer' loosely interchangeable during the Renaissance, when most princes employed such a professional at their courts. But whereas astronomy was concerned only with prediction of positions as such, from the 'natural' character attributed to the related bodies further 'natural' consequences could be inferred. Here the 'physicist' stepped in: the man who was able to interpret the relations of all natural bodies in terms of the Aristotelian 'elements'—earth, fire, water, and air; the closely related medical 'humours'—phlegm, blood, yellow bile, and black bile; and the 'temperaments' of individual men consequent upon their particular humoral balance. The astronomer was concerned only with objects above the sphere of the Moon; the physicist only with 'sublunary' bodies. Since the former were supposed to be devoid of the elements (except the 'ether', specific to themselves) the link between the two regions inevitably involved a highly hypothetical 'influence', the nature of which greatly exercised the minds of Renaissance thinkers. The astrologer as such was not necessarily concerned with the nature or means of transmission of this 'influence', only with its assumed effects. His business was to estimate the balance of influences from the observed data, the ultimate aim being to foretell the course of nature—storm, flood, harvest, plague; as well as individual human destiny. Once more it must be emphasised that the complete sceptic about the possibility of this was rare indeed. Where disagreement broke out—sometimes reaching a high degree of academic acrimony or, where ecclesiastical authority was involved, judgment of life or death—was the degree to which the celestial influences *determined* the fate of the human soul. A little further analysis of the issues at stake throughout the whole Renaissance period will not only help to clarify one of the major relations between science and society at that time, but may remove some of the more naive conceptions of science as it is today.

To count as a 'science' today a corpus of knowledge must before everything be based on 'publicly' observable data—what are loosely called 'observable facts'. The basis of all astrological prediction was (and of course is) the horoscope. This combination of Greek words precisely describes its nature—a 'view' of the 'time'. There is nothing in a horoscope as such that cannot be *observed* by anyone trained to use certain instruments. The data (set out on a characteristic chart) comprise the angular distances of the visible planets from each other ('aspects'), their positions with respect to the zodiac, and the sign of the zodiac rising on the eastern horizon ('ascendant'): this constitutes the 'view'; the 'time' may be of conception, birth, a planetary conjunction, eclipse, laying of foundation stone, setting out on a voyage, joining of battle, etc. Since a few minutes one way or the other in the time might make a very considerable difference in the predictions derived from the horoscope these various components would not normally be actually observed but calculated from the single observation of the 'time'. Strictly speaking, since there were no really reliable mechanical clocks until late in the seventeenth century, even the time would not be actually observed but arrived at by observation of some standard star and subsequent calculation of the solar time by reference to tables. It was for this reason, as the title of the printed work suggests—*Tabulae Directionum . . . navitatibus multum utiles*—that Regiomontanus gave priority to the publication of tables of the positions of the Sun, Moon, and other planets. It is significant that the first printed edition was edited with an introduction by Bilibald Pirckheimer, a leading 'patrician' humanist of Nürnberg, friend and patron of Albrecht Dürer.

Without going into the technical niceties of interpreting the 'face of the heavens' at the accurately established 'time' it may be well to sketch in the barest outline the significance of the principal features of the horoscope. Of these, unless there was any unusual occurrence such as an eclipse or planetary conjunction, the most important were the 'signs' of the zodiac in which the Sun and Moon happened to be situated, and the 'ascendant' itself (p. 36). The zodiac is a belt bounded by great circles 8° north and south of the apparent circular path of the Sun through the heavens (ecliptic); the Sun, Moon, and other planets are never seen beyond these limits. The exact nature of the 'signs' has been a source of controversy that has come down to our own time. They may be regarded either as constellations having the names of eleven animals and the Scales or as those equal divisions (30°) of the zodiac where the constellations were at a certain epoch: this ambiguity is the consequence of the so-called 'precession of the equinoxes'.

Thus far the horoscope is seen to be merely a visual summary of astronomical 'facts' calculated by the best available mathematical procedures from observations made with the best available instruments. Though the whole exercise may be considered a futile waste of time and ingenuity, there is nothing 'unscientific' about it. For no science consists merely of 'facts'; it is the possibility of explaining or predicting other kinds of facts that turns a 'collection' into a 'system', and here *hypothesis* must play its part. It is the—to most of us—wholly arbitrary and far-fetched interpretative hypotheses that render astrology, not necessarily false—of this we can never be *sure*—but sterile. These hypotheses refer to the specific influence of each planet, the general character of each sign in which the planet may be found, and the mutual interaction of the planets as a *consequence* of their 'aspects'. One other factor is generally involved— the most removed from scientific objectivity—this is the arbitrary division of the sky into regions characterised, not by an 'influence', but by the nature of the circumstances—health, wealth, journeys, etc.—in respect of which the 'influence' may act. It would hardly be necessary even to mention this factor were it not the case that the system of these so-called 'houses' most commonly employed in the sixteenth century was devised by Regiomontanus himself—the man held by his contemporaries to have been most influential in 'rescuing' astronomy from the 'stagnation' of the Middle Ages.

In modern science the value of a hypothesis depends to a large extent on the possibility of showing, by recourse to new observation or experiment, that its necessary consequences are false. The most dramatic of such tests of Renaissance astrology came in 1524 when something approaching a second deluge was forecast by Ioannes Stoeffler in consequence of a concentration of planets in the sign of the Fishes. The absurdity of such an inference is patent; nevertheless large numbers of German people of all walks of life went up with their belongings on to the high mountains in the hope of escaping the wrath to come. Of course there was no 'wrath' except against the prophet. Now Stoeffler was no idle mystic living in a dream world of incantations but the fabricator of a famous clock in the cathedral at Strassburg (replaced by the present one in 1574) and of numerous instruments; his book on the design and use of the astrolabe was hardly surpassed throughout the sixteenth century.

There were of course many other instances of falsification, but there were also many successes, as there were bound to be. But before we write off the 'science' of astrology as wholly bogus it would be well to recall, what now seems to be forgotten, that only forty years ago leading astronomers were laying 'astronomical' odds against the *possibility* of planetary systems anywhere else in the universe. Fortunately for the writers of SF 'scientific' fashions change even more radically than do those of female adornment.

Though astrology will raise its protean head again more than once in our narrative we may leave it for the present, and take a look at those activities of Regiomontanus which display a changing relationship between science and society, whether or not motivated by the prevailing obsession for astrology.

Regiomontanus, it will be recalled, taught Greek in Italy before he had established a reputation as a mathematician. It is not surprising then to find that one of his most important enterprises was to draw up a 'sales catalogue' of forthcoming critical editions of the broadly mathematical Greek classics without which, as he alone seems to have recognised, solid progress in mathematics and astronomy was unlikely. Such a task was evidently beyond the resources of a single scholar; but it is significant of the Nürnberg society that a young patrician, Bernhard Walther, was prepared to advance the necessary capital and provide personal assistance; so the first 'technical press' in Europe came into being. But this was not all; a permanent observatory was established whose observations could be published with a minimum of delay by means of this private press.

Owing to the untimely death of Regiomontanus the fruits of this enterprise were meagre indeed; but for our purpose it is the *idea* and the circumstances in which it was put into effect that are significant rather than the expected fruits.

If the market for improved astrological wares may have been the principal motive for Regiomontanus' decision to settle in Nürnberg, it was certainly not the only social factor involved in the German mathematical Renaissance. It was formerly widely held that commerce with the East was the chief motive, and that it was the tables of Regiomontanus, based on more accurate observations with more modern instruments and calculated by more advanced mathematical techniques, that enabled Christopher Columbus to discover America. The general rejection of this view will be dealt with in a later chapter; but to put the matter in proper historical perspective it may be said at once that in each of these 'new' factors Regiomontanus lagged behind the Persian astronomers of the thirteenth and Mongol observatories of the early fifteenth centuries: for his *equal* in respect of instruments and originality of approach it is not necessary to look further afield than to the Augustinian Houses we have already taken note of. It would, however, be equally misleading to ignore the surprising interest—for a city so far removed from any outlet to the sea—in the problems of oceanic navigation; but here also the evidence relates entirely to the decades following Regiomontanus' death.

The circumstances of his death are actually related to the last of the social factors involved in the cultivation of astronomy. Since, despite earlier discussion, the civil and ecclesiastical calendar (Julian) was about ten days behind the solar, Pope Sixtus IV called Regiomontanus to Rome to advise on the establishment of a revised chronology. Soon after his arrival in Rome he died—by poison, according to rumour—so reform was delayed for more than a century, when another German, Christoph Clavius of Bamberg, played the principal part in setting up the present system of time-reckoning named after the pope, Gregory XIII, at whose instigation the reform was carried through.

In the eminent part played by the society of Nürnberg we cannot of course ignore the fact that it was there one of the most powerful minds of the Renaissance, that of Albrecht Dürer, came to maturity; but he was only five years old at the time of Regiomontanus' death. The other major field of German culture which became involved with contemporary science had its centres at 150–200 miles from Nürnberg—Chemnitz in the north-east, East Tirol and Kärnthen (Carinthia) southwards: this was *Bergbau* and *Bergkunst*—mining and metallurgy.

The relation between the mining industry and the science of the

Renaissance was wholly different from any that we have so far considered. Whereas the cultivation of astrology was conditioned by a sophisticated mathematical science, the practice of mining was 'scientific' only by virtue of its dependence on experience. Its procedures lacked any agreed theoretical basis; they were controlled by the 'pragmatic sanction' of delivering the goods. There were no learned treatises; practice was based on tradition—almost wholly oral or perhaps contained in closely guarded manuscripts that have disappeared, leaving only fragments. The only record of progress in the art was the swelling of the ledgers of large-scale operators, who in the second half of the fifteenth century rapidly replaced or absorbed the scattered peasant units. It is no exaggeration to call this a mining 'explosion', since when the greatest of these entrepreneur-bankers, Joseph Fugger, died in 1525 an Imperial mandate estimated the German output as worth two million gold gulden, and claimed that a hundred thousand people were involved in it. Such 'statistics' of population are notoriously unreliable; but more trust can probably be placed on those where a 'cash-nexus' is involved; the art of banking, book-keeping—and taxation—was by then highly developed. The political consequences of this German near-monopoly of gold, silver, copper and tin were of course profound. The details do not concern us; but we can hardly doubt that the consequent establishment of German banking houses in the leading cities of Northern Italy assisted the cross-fertilisation of ideas already noticed. The greatly increased production of copper and tin facilitated the 'improvement' of ordnance; before the end of the fifteenth century mobile artillery as well as the earlier siege trains had become an indispensable component of national armies.

If the impact of scientific theory on mining and metallurgy was, and for a great part of the sixteenth century remained, virtually nil, the Northern Renaissance was far spent before the converse relationship became manifest in the figures of Paracelsus and Georgius Agricola (see p. 137), and which owing to its very complex nature must be deferred for later consideration. Nevertheless it is significant that the two 'artists' who were most concerned with the 'nature of things' both served their apprenticeship in circles devoted to metal-work—Leonardo da Vinci in the *bottega* of Andrea Verrocchio, where problems of metal-casting were very prominent, and Dürer in goldsmithery. The mingling of metals and regard for the properties of the resulting alloys may be traced back almost to the dawn of civilisation, giving rise to the useful if rather misleading chronological divisions of Copper, Bronze and Iron Ages. More important for weapons of attack and defence is brass whose use was well established among the Romans. Brass is peculiar in being composed of the very well-known copper and a metal, zinc, that was almost certainly unknown as *such* until the seventeenth century, the alloy being produced by mixing the well-understood source of copper (e.g. malachite) with an 'earth', calamine (zinc carbonate), that greatly altered the copper. Failure to recognise the presence in calamine of the metal zinc in contrast to that of tin in bronze is of course due to the high volatility of zinc which, if separated from the ore, 'boils off' and is oxidised to a white powder whose presence was often observed. If anyone (at least in Europe) could be credited with the recognition of a 'latent' metal it would have been Paracelsus, whose interest in metals must have been directly founded on his early and intimate experience in the Fugger mines of Carinthia, where his father acted as surgeon to the miners; but his use of the actual word *zinzi* is no guarantee that he had a proper comprehension of what was involved.

In concluding this sketch of the Northern Renaissance it must be emphasised that the absence of reference to any European countries other than Germany and the neighbouring kingdoms can be justified only in respect of *science*. Nowhere else (with the possible exception of the Iberian Peninsula, see p. 70) does any major figure stand out revealing any serious concern with new, or renewed, modes of thought relating to natural knowledge. In Paris as late as 1516 a Scottish master, George Lokert, commented on the decay of interest, *outside Italy*, in the works of Jean Buridan and Albert of Saxony, who two centuries earlier had brought a critical insight to bear on Aristotelian natural philosophy. But in respect of the literary Renaissance, though it is still true that the Italian revival first found an echo in Germany, it would be a grave mistake to assume that classical Latin and even Greek were not being cultivated at all widely before the end of the *quattrocento*.

3

'MEN CAN DO ALL THINGS IF THEY WILL'

The actual words are Alberti's but echoes of them occur in the writings of Dürer, Leonardo da Vinci, Paracelsus and other great Renaissance figures. There could hardly be a more succinct indication of the spirit that motivated the transition to modern science. But since the time when Burckhardt quoted them at the conclusion of his famous evocation of Alberti the unconscious complexity of Burckhardt's notion of the 'discovery of Nature and Man' has been revealed; it will be the concern of this chapter to investigate the nature of this complexity.

Here, as elsewhere, it may be best to start with the man who in the popular imagination is probably still looked upon as the pioneer; the man who may be regarded as nearly, if not quite, 'the first modern scientist', appearing almost out of the blue and barely appreciated by his contemporaries. Leonardo's almost universal genius and pre-eminent achievement as painter and draughtsman are, of course, beyond question; but this in itself constitutes no claim to *scientific* genius. Such a claim cannot be founded on any actual achievement, of which indeed there was very little, but on the astounding range of significant ideas revealed in his sketches and notebooks. In the assessment of these, however, too little attention had until comparatively recent times been paid to the achievements of his contemporaries or forbears, which he was able to use as a springboard into his own more comprehensive world-view.

Although almost solely in the realm of painting are any concrete *achievements* of Leonardo to be found, it was of the highest significance for the development of his wide-ranging genius that at the age of thirteen, when boys of intellectual promise would be about to enter a university, Leonardo entered the *bottega* (workshop) of

Andrea Verrocchio, where he was not only taught the actual technique of painting but also that of metalwork. More important, he
would almost certainly have had a closer view of the anatomy of
the human body than at any contemporary medical school; and,
perhaps most important of all, he would have heard such visitors
as Toscanelli discussing the technical problems of the day. There
Leonardo remained a dozen years.

Among these 'technical problems' none would have been of more
fundamental importance than that of 'perspective'; for although
arising out of the painters' problem of representing solid objects
on a plane surface, the new 'image' of space which emerged at that
level, sometimes by way of analogy, sometimes in sharp contrast,
was related to the general world-views of the thinkers of the Renaissance; and these in turn had a profound if subtle influence on
Renaissance society.

Leonardo, as Kenneth Clarke has pointed out, must have built
on the highly technical treatise of Piero della Francesca (p. 49),
which, though entitled *De Prospectiva Pingendi*, went far beyond
painting in its method. Piero in turn may have learnt the elements
of the subject from Alberti; and we can now see that, though for
purposes of 'popularisation' Alberti had produced the first 'primer'
of the subject, it was in fact Brunelleschi who broke away from the
medieval discipline of *Perspectiva* and the traditional perspective
dodges employed by Greek artists. For it was not the *problem* that
was new: the Greeks in fact were so adept at creating the illusion of
a solid on a plane surface as to call down the censure of Plato for
their use of such 'deception'. Byzantine mosaics (e.g. the *Good
Shepherd* in the mausoleum of Galla Placidia at Ravenna) could
achieve a high degree of naturalism. During the High Middle Ages
there was certainly a general failure to achieve a natural appearance
of solid objects, but at least a century before Brunelleschi the use of
foreshortening was again becoming more effective. The men of the
Renaissance themselves looked back to Giotto (d. 1366) as the one
who had 'restored art'. Finally Jan van Eyck's *Marriage of Giovanni
(?) Arnolfini* (1434) displays technical mastery in no way inferior
to contemporary Italian achievement. All these were achievements
of the highest *art*; Brunelleschi probably was the first to demonstrate
that complete understanding of perspective can be gained only by
treating it as a problem of *science*; that is, by analysing the act of
vision as it actually occurs when directed at a natural object. On the
assumption ('proved' by Euclid) that light follows a rectilinear path
(the *direction* being immaterial) only the angle subtended at the
retina is significant: foreshortening of bounding surfaces is no longer
a 'dodge' but a geometrical necessity; and the necessary degree of
'foreshortening' can be worked out precisely (as it was in Piero della

Francesca's *Flagellation of Christ*) by means of geometry, using as a standard of length (the 'module') an object whose size is intuitively known, such as the height of a man.

Though this attempt to grapple with man's intuition of space was the most scientifically important feature of Alberti's work *On Painting*, a more immediate effect on contemporary society, since it resulted in works more easily accessible to the public gaze, was his physical world-view. His master-notion is expressed in the words 'Serious painters . . . will put as much serious study and work into remembering what they take from nature as they do in discovering it'. Truth to (ideal) nature is the final test; though in constructing his *istoria* (that is, not a mere copy) the painter must select what is most beautiful. *What* he will always find in nature is dealt with more specifically in Alberti's work on architecture. Everywhere in nature is to be found geometrical harmony; above all, patterns based on the circle—recognised by both Plato and Aristotle as the most 'perfect' figure—and to a lesser extent the square. Furthermore the consonance of Man and Nature—and this is one of the basic aspects of what is being examined in this chapter—is revealed by the fact that the ideal human form fits without distortion into a square and circle. This harmonious relationship was first stated by the Roman architect Vitruvius; it is exemplified in drawings and printed figures by a succession of Renaissance architects during the two centuries and more succeeding Alberti's work. The corollary for Alberti and for these architects was that the proportions of the 'noblest of all buildings', that of Christian 'temples' (Vitruvian in contrast to the medieval rectangular basilica), must follow the same model. The details of architectural development do not concern us, but the most perfect realisation of the form (sketched but never constructed by Leonardo) is S*anta Maria della Consolazione* that comes into view almost in isolation as the traveller leaves the hill-town of Todi on the road down into the valley. It was begun in 1504 by Cola da Caprarola, probably following a design by Bramante.

The other aspect of Alberti's physical thought that should occupy our attention at this stage is his emphasis on movement, not only in respect of the greater emphasis on action in the Renaissance *istoria*, but specifically in respect of light and the 'machinery' of animate beings. 'Colours are borne by rays', he asserts, hence the change of the apparent colour of distant objects, whose rays become 'tired' and whose outlines hazy as a consequence of longer travel through 'humid air'. The juxtaposition of appropriate colours thus becomes a significant aid to the representation of aerial perspective. No picture painted by Alberti survives, but his late contemporary and almost certain associate, Piero della Francesca, was to demonstrate this device with consummate artistry. The notion of 'moving' light goes back of course to Plato and Euclid, but theirs is a wholly subjective view as of an insect's antennae *seeking* the object. The modern view (founded indeed on no decisive evidence) of vision being mediated by rays *from* the object appears first in ibn al-Haitham (eleventh century) who first gave a largely correct view of the mechanism, based on dissection of the eye. Though, as has been noted already, the *direction* of propagation is wholly irrelevant to correct geometrical inference the objective origin of rays accounts much more plausibly for the effects of distance. As in the case of colour contrast, so in the application of chromatic and boundary *chiaroscuro* Alberti's notions received adequate development only later, notably by Leonardo, whose theoretical treatment was probably his most important contribution to the 'science' of painting.

Alberti's remark that 'it would be useful to isolate (*allegare*) each bone of the animal and on this add its muscles then clothe all of it with its flesh' ('skin' would be more accurate in view of the term 'flesh' being generally applied to muscle) shows that in their aim to represent *living* animals accurately the artists had far greater insight into the 'science' of anatomy than seems at all to have been the case in the contemporary medical schools. For the latter, anatomy was much more concerned with the hollow viscera—gut, heart, liver, brain, etc.—in which various 'spirits' were supposed to be generated; anything approaching an adequate representation of muscles was rare and late in appearance, and medieval physicians paid scant attention to this aspect of the admirable anatomical descriptions of their master, Galen. For the artist's portrayal of the human body in *action* (p. 107)—not necessarily *motion*—a knowledge not only of the position but also of the changes of size of individual muscles was essential for an understanding of the material mechanism effecting the relative motions or stresses of *parts* of the body. Such mechanisms become intelligible only when the shortening of individual muscles can be seen in relation to the underlying rigid skeletal framework; in other words, anatomies must be carried out 'backwards' as well as 'forwards'. This is precisely what Alberti is saying; nor is there any reasonable doubt that the young Leonardo saw such procedures in the *bottega* of Verrocchio. The Renaissance obsession with the *human* form accounts for the relatively slow emergence of even the data for *comparative* anatomy: the fact that the drawings of monkeys and hounds by the otherwise *gothic*-styled painter, Pisanello, are outstandingly realistic in respect of the musculature of the limbs should give us pause when we are tempted to generalise about the *Renaissance* 'discovery of the world and of man'.

Since Alberti died when Leonardo was only sixteen his influence, though clearly of the greatest importance, must have been largely indirect; the same may be said of Piero della Francesca's. But in respect of Leonardo's remarkable versatility in the design of machines both for peace and for war, we have the evidence of a manuscript with numerous marginal glosses in his unmistakable hand. This MS is now confidently attributed to the Sienese artist Francesco di Giorgio Martini, whose relation to Leonardo will be considered later (p. 49). Finally, there is no doubt at all that in respect of mathematics Leonardo owed a great deal to Luca Pacioli (a younger contemporary of Piero della Francesca at Borgo San Sepolcro) without whose aid he would probably have been unable to apply the methods he read about with difficulty in the works of Euclid and Archimedes, neither of whom was then available in Italian translation.

Now it can hardly be a mere coincidence that all three of these 'precursors' were in the employment of, or in intimate association with, Federigo da Montefeltro, Count, and later Duke, of Urbino. It will repay us to look rather closely at the character of this remarkable man and the miniature 'Renaissance society' that he brought into being.

On 10 June 1468, when Leonardo da Vinci was sixteen years old, Federigo da Montefeltro signed letters patent appointing Luciano Laurana 'Engineer and Chief of all the masters' working on the palace of Urbino which he had decided to build 'as a residence beautiful and worthy, such as is fitting the condition and praiseworthy fame of our forbears and ourselves'. Federigo was the natural son of Count Guidantonio and in 1444 had succeeded his half-brother Oddantonio, whose assassination was formerly believed to have been symbolised in Piero della Francesca's *Flagellation of Christ*. Another document reveals that Laurana was already in a position of authority a year earlier, and there is little doubt that work was already in progress on the integration of a number of buildings, one at least of which had formed the residence of earlier Counts of Urbino. The importance of the patent of 1468 is that its preamble reflects the cultural aim of a man who among Renaissance princes was the counterpart of Alberti among the patrician families.

There were many princes anxious to vie with each other in the magnificence of their palaces and the scale of their patronage of the creative artists and scholars in Italy and elsewhere, but their own accomplishments, put in the most generous terms, were restricted to military conquest. To this generalisation there were one or two exceptions, such as Lorenzo de' Medici, an accomplished poet, who was personally involved in the activities of the so-called 'Platonic Academy' (p. 152) founded by his grandfather, Cosimo. But in contrast to Lorenzo, shrewd banker, political juggler, and dilettante, Federigo had risen to fame on the reputation of his victorious exploits as *condottiere*. And when firmly established as a wise and generous but still formidable ruler of the small hill-city of Urbino it was not to *belles lettres* that he turned for recreation but to the austerities of theology and mathematics, and the ingenuities of mechanical contrivances for peace and war. Perhaps it was the influence of Alberti, whose most famous 'monument' is the 'tempio' to Isotta Malatesta in the rival and not far distant city of Rimini, that moved Federigo to hail architecture as an 'art of high science and genius, since it is founded in the art of arithmetic and geometry that are the chief among the seven liberal arts, being of the highest level of certainty [*in primo gradu certitudinis*, the patent being otherwise in Italian]'. However that may be, it is no exaggeration to say that out of the revised plans sketched for him by Laurana some time before 1468 grew up not only the loveliest secular building in *quattrocento* Italy but a shrine in which the ideal of embodying natural harmonies in human structures could be realised.

In 1472 Laurano left Urbino, but there is evidence to show that a great deal of the work—mainly decorative, indeed, but integral to the whole conception—though carried out after his departure was nevertheless inspired by his grandiose plan. Much of the decorative work and a few major architectural structures bear, however, the stamp of an individual master of the front rank; this was Francesco di Giorgio Martini, brought from Siena to work in the palace before Laurana's departure. The additions justify the postulation of a third phase in the architectural history of the palace. There was also a fourth during the rule of Federigo's son, Guido-baldo, and of the members of the succeeding della Rovere family; but no fundamental changes were made, and those decorative features that cast a light on the vigorous intellectual activity characteristic of the later years of Federigo's reign had already been completed under his own supervision.

Though subjected to depredation by later occupants sufficient evidence remains in the extant palazzo to enable a plausible reconstruction to be made of that part of the building where the blend of humanistic and technical activities fostered by Duke Federigo had its centre. Between the two towers that are the most striking external feature of the building is a tier of three loggias, one of which gives access to a vestibule out of which doors lead into the Duke's private apartments, including most significantly his *studiolo*. Before describing the decoration of this room, which gives the key to its importance for our enquiry, it is relevant to add that on the floor immediately below are two small rooms of which one is still recognisable as a superb example of the domestic chapel of the period, and the other corresponds to what was recorded as being a temple to the Muses. This juxtaposition of Christian and pagan worship is no isolated instance in the art and custom of the *quattrocento*, and ought always to be borne in mind before making sweeping generalisations (according to individual prejudice) of the 'paganism' or 'ecclesiastical preoccupation' of the Italian Renaissance. Also on the ground floor, but in the central part of the building, was the great library, most of whose priceless contents were rifled in 1502 by Cesare Borgia. Further reference to this outstanding collection will be made in more than one connection.

The conjectural reconstruction of the *studiolo* from both the extant mural *intarsie* and the identifiable scattered fragments leads to the conclusion that it was meant to provide an image of scientific activity effected by the representation of very varied concrete objects: books, musical instruments, a teaching armilla, the mechanism of a weight-driven striking clock, arms and armour, a reading lamp with a book leaning against it, a pipe organ and—divested of his accoutrements—the Duke himself. An essential part of the 'concept' is the signs of disarray—books and instruments withdrawn from their cupboards and left lying about: this was no static conventional 'monument' to a great man, but the representation of a workroom where he could indulge the insatiable desire for an active life of the mind and senses. The studious side of his nature was emphasised in the famous portrait of himself, still armour-clad, reading a massive volume while his patient little son Guidobaldo stands beside him. This almost certainly occupied a central position among the painted portraits of famous men including such 'scientific' figures as Ptolemy, Euclid, Bessarion, Hippocrates, and Pietro d'Abano (p. 33) besides the more conventional figures. Notable also is the presence of his own teacher, Vittorino da Feltre, from whom at the court of the Gonzagas at Mantua he had imbibed that idea of harmonious development of human faculties that was perhaps the most strikingly novel characteristic of the Renaissance.

Such, we may believe, was both the daily vision and the memorial of this extraordinary man; a creation both embodying and illustrating the prevailing 'science'. The apportionment of the creative acts between those masters who were known to be associated with it is a problem unlikely ever to be solved. For the master-plan Rotondi thinks Bramante to be the most likely candidate. In respect of the *intarsie* we have to distinguish between their design and fabrication. Though most of the panels may have been worked in Florence under the supervision of Sandro Botticelli, it is unlikely that they were all designed by him. The panel in which a lovely landscape vista is glimpsed between the capitals surrounding a courtyard (or spacious loggia) almost certainly derives from Francesco di Giorgio.

To discover how this hive of creative activity stood in relation to those who were employed there we have to turn to literary sources, one writer in particular having been well placed to lay bare the web that connected so many of the leading creative minds with Duke Federigo and with each other: this was the Franciscan friar, Luca Pacioli, whose *Compendium of Arithmetic, Geometry, Proportion and Proportionality* was printed in Venice in 1494, mainly in Italian but with a few sections repeated in Latin—the first *systematic* printed work on mathematics in any vernacular. In the dedication to Guidobaldo, Pacioli calls to mind many of the great contemporary artists, among whom he specially emphasises Piero della Francesca ('my countryman' [*conterraneus*]—they were both citizens of Borgo San Sepolcro) 'whose book on painting and linear perspective not unworthily found a place among very many others relating to all subjects in your splendid library'. Piero's *De Prospectiva Pingendi* was in fact dedicated to Federigo, his later work on the regular polyhedra to Guidobaldo.

Pacioli, it will be noted, rather strangely omits to mention Francesco di Giorgio, and even more surprisingly his own pupil, Leonardo da Vinci. But though the book was printed only in 1494 the scene he depicts in the dedication must be that of a much earlier date since Piero, whom he calls the 'greatest (*monarcha*) painter of our time' had given up painting by 1478 and in 1494 was already dead. Printing was still something of a novelty, and it was no uncommon thing for a work to be printed only after the importance of the matter contained in it had been demonstrated by its distribution in manuscript. Pacioli's close association with Leonardo occurred when they were both living in Milan. It was there, in a book subsequently printed in Venice in 1509 under the title *Divine Proportion*, that Pacioli paid his master, Piero, the dubious compliment of passing off as his own most of Piero's work on the regular polyhedra. In the dedication Pacioli refers to Lodovico Sforza and 'his Vinci' who had drawn the diagrams of the polyhedra (including the very difficult star-polyhedra) for the book.

Of this aspect of the Urbino network there has, thanks to Pacioli, never been any doubt. The full significance of Francesco di Giorgio, on the other hand, has come to light only in fairly recent years; not only is his name absent from the writings of most of his contemporaries but the manuscripts on which almost all of our knowledge of his creative imagination is based bear no certain clue to their authorship, which has however been established by modern scholars beyond all reasonable doubt.

Leonardo met Francesco in Pavia (the latter's birthplace and scene of some of his most important work) in 1490 and, as noted above (p. 46), possessed and annotated the copy of Francesco's principal work still extant. On looking into the magnificent fascimile now available one can readily appreciate Leonardo's enthusiasm: the ingenious devices for the transmission of mechanical effort— sprockets, crown wheels worm-drives, 'gear boxes'—are displayed with a clarity rivalling Leonardo's own sketches. Perhaps even more significant is Francesco's admission that, though the great wealth of reference books in the library at Urbino had been of inestimable value in the ordering of his material, he had preferred to base his statements on his own findings rather than on those of any classical author, some of whose works he had found to be un-reliable. And when it came to military science he went further, actually resorting to quantitative trials in determining the best proportions for offensive instruments (powder, balls) on the one hand and defensive works on the other. It would be wrong to infer that Leonardo relied on Francesco for inspiration; despite the near impossibility of dating most of his sketches there is no reason to assume that the most ingenious post-dated his study of Francesco's work. What can be inferred is that Leonardo was in point of mechani-cal invention no solitary genius far in advance of his time. Indeed the evidence goes to show that Francesco di Giorgio's was only the most ingenious and 'scientific' of a great number of sketchbooks of mechanical devices of very varying merit. Less erudite and mathe-matical than Alberti, his imagination was less restricted by classical models. Though no more 'literate' than Francesco, Leonardo combined, as perhaps none had before him, the sense of good design and empirical approach characteristic of Francesco, the mathematical control of Alberti and Pacioli, and the deep physical insight that, as we shall see, he owed to no predecessor.

In no aspect of the aim at mastery of the environment by human agency did Leonardo transcend his contemporaries more than in the study of *motion*. In his realisation of the importance of static pressure of water he showed no great superiority to Francesco di Giorgio; there is evidence that the latter had at Urbino provided for both the drainage of the immense stable area and snow disposal from the raised garden, for whose fountain he installed a high-level cistern. Doubt even has been cast on Leonardo's priority in the invention of the mitral lock gate. The theory of hydrostatic pressure had been established axiomatically by Archimedes to whom Leon-ardo was constantly referring; but so far as Western Europe is concerned *motion* was until the early fourteenth century interpreted in terms of the confused theory of Aristotle. The medieval followers of William of Ockham had introduced the important distinction between kinematics (*how* bodies move) and dynamics (*why* and *with what effect*). The former was conceived almost wholly (it was Galileo who removed the 'almost') as an abstract relation between space and time. To the passionate artist, Leonardo, such a 'disembodied' motion would have had little appeal. To the concept of *impetus*, by which the medieval 'physicists' had hoped to remove the supposed necessity for a *persistent* mover, that of *forza*, whose manifestations Leonardo saw everywhere in nature, owed much. Like *impetus* it placed the 'cause' of motion *within* the moving body itself. In so far as it informed not only rushing water but living creatures it resembled more nearly the later *vis-viva* ('living force') or 'energy'. His studies of turbulent motion are of outstanding importance for our present concern. When we recall that the prime mover of medieval technology was largely water we can appreciate that the delineation of the water below the weir or sluice-gate was a challenge as much to the engineer as to the artist: indeed the former was in the ascendency, for Leonardo's sketches are those of the 'morphologist' concerned with what is *actually* happening rather than what to the untrained unconcentrated vision *appears* to be happening. But the consequent combination of the instantaneous image with the suggestion of rapid change was an astounding achievement.

It is, however, in his obsession with the problem of flight that Leonardo's recognition of the necessary condition for progress in natural knowledge most clearly stands out—the distinction between *experience* and *experiment*, foreshadowed indeed on a limited scale by his medieval forerunner, Roger Bacon. The nature of his problem precluded until the coming of the wind-tunnel the wholly artificial and controlled 'experience' that constitutes 'experiment'; but within this necessary limitation Leonardo's systematic observations of wing-structure, wing-beat flight, soaring flight correlated with up-currents, use of the tail for balance in stationary flight and as a 'landing-flap', and many other factors reveal a remarkable appreciation of experimental methodology. Where significant characteristics were concerned rather than actual flight he was able to make experiments, as in the determination of centre of gravity of a bird in various flight-postures. Here also is his most advanced appreciation of theory—the control of centre of 'resistance' (air pressure) in relation to centre of gravity as a means to attain equilibrium, a problem whose solution involved the intuitive recognition of the *moment of a couple*, to give it the modern term.

Nowhere in Leonardo's work does his character as a representative of his age stand out more clearly than in the following passage: 'A bird is an instrument working according to mathematical law, an instrument which it is *in the capacity of man to reproduce* with all its movements though not with a corresponding degree of strength, for it [*sc.* the human machine] is deficient in maintaining equilibrium. We may therefore say that such an instrument constructed by man is lacking in nothing except the life (*spirito*) of the bird and this life must needs be supplied from that of man.' *Men can do all things if they will*—that is, if they base their plan on experience subject to mathematical law and control it by experiment; this alone is the way to *certain* knowledge. But there is a catch here that Leonardo didn't fully appreciate and his nineteenth-century biographers ignored; this is the presence in nature of *spirito* and *forza*. Leonardo is honest enough to admit that these are inescapable aspects of actual *experience*, and as such he finds it necessary to employ them in his 'explanations' of nature; but he is silent about their complete independence of mathematical law. Fortunately his practice was more catholic than his rhetorical insistence on mathematical formulation would have permitted, otherwise there would have been no anatomical studies—perhaps the greatest of all his scientific *achievements*.

On the other hand this failure to adhere to the ideal of mathematisation may have been responsible for his failure to bring to any degree of fruition the astonishing insight he gained into the mechanics of flight. For despite the inherent futility of the analogy, his obsession with wing-flapping as the means of human flight can hardly have had any other cause than the belief in the necessity of human 'life' to inform the mechanical copy. It is characteristic of the extraordinary inconsistency of Renaissance man that it was Girolamo Cardano, far more credulous of the active power of magic than was Leonardo, who dismissed the whole project with the devastating comment: 'Leonardo also attempted to fly, but misfortune befell him from it. He was a great painter.'

We are here face to face with a fundamental question in regard to the nature of science—is the 'science' of an epoch to be evaluated solely in terms of its quantitative and mathematical form? If the level of scientific achievement is to be measured by the degree of prediction and control it makes possible, then the answer is unquestionably, yes. It is often argued that the importance of the Renaissance in the history of science is not the discovery of new facts, of which there was very little except at the purely descriptive level (p. 73), but the recognition that progress would be possible only to the extent that knowledge of nature could be given the precision necessary for it to be expressed in mathematical relationships; a picturesque but misleading way of saying this has been to point to the 'triumph of Platonism' over Aristotle, or in the subsequently revised form of the triumph of Archimedes over both. What has already been said about the *quattrocento* artists and their heroic successor, Leonardo da Vinci, shows that this view embodies an important degree of truth. But to a modern student of science, unfamiliar with the cultural atmosphere of the Renaissance, it can be highly misleading; it can even be used as evidence for the fashionable view that it was all part of the 'humanist' campaign to bring light into the dark places of the mind by throwing off the authority of the 'Church'. As has been hinted above (p. 44) the *geometrical* harmony embodying the square and circle had already replaced the much more complex gothic forms in the design of churches before Leonardo was born. But the notion went deeper than this, and in the *cinquecento* was to go deeper still, when it became linked with the art even more closely connected with 'science', namely music. The absence in Leonardo's writings of any concern with the vigorous prosecution of new musical forms then prevalent has often been remarked.

The ultimate source of this ideology, deeply ingrained within the Western spirit, was the words of Holy Writ 'By measure and number and weight Thou hast ordered all things' on which, as Pacioli reminded his readers, St Augustine had some weighty things to say. Augustine's remarks indicate that he was interpreting *measure* as *degree* rather than as number; but it was in the more mathematical form of *proportion* that Boethius (the 'last Roman heir of Greek philosophy') passed it on to the Middle Ages, in his theory of music. It is important to recall the profound (and almost certainly restricting) influence on Western science of the absence in Greek mathematics of the concept of 'real number', the nearest approach being the ill-defined notion of 'magnitude' ($\mu\acute{\epsilon}\gamma\epsilon\theta\sigma s$). This was no obstacle in the development of geometry, expressed as this was in terms of the lengths of line-segments and their ratios together with those 'magnitudes' (square, cube, etc.) that were products of the 'same magnitude'. But of a product or ratio of two *different* kinds of magnitude (e.g. the 'moment' of a lever or the ratio of length to time—velocity) they had no conception. Galileo's elephantine effort to express in words the appropriateness of what we regard as his fundamental equation of free fall ($v = u + at$) may have been part of the long shadow cast by this obstacle. Such 'mixed ratios' were frowned upon in school arithmetic even in the present century.

Throughout the sixteenth century at least in the universities, there remained the residue of the medieval conception of the theory of numbers expressed as ratios (*proportiones*) and as the relation between *proportiones* known (from Boethius) as *proportionalitas*. From the time of Pythagoras (sixth century B.C.), if we can trust the earliest fragmentary records, this notion was linked with the world of sense by the 'experimental' demonstration that the consonances natural to the human ear were produced by dividing the sounding string of the lyre in the ratios 1:2, 2:3, 3:4 (octave, fifth, and fourth). Since these four numbers—1, 2, 3, 4—were later found to be sufficient to express also the composite consonances-octave plus fifth (1:2:3), and double octave (1:2:4)—they were held to be in some way sufficient to express the harmony of the world. This embraced both the cosmic (planetary) intelligences, whose music was bodied forth by the rotation of the spheres that carried them round in their eternal motions, and the human soul. The solution for Plato lay in adding to the fundamental numbers the squares and cubes (4 being already cared for as the square of 2) thus obtaining two geometrical progressions (1:2:4:8) and (1:3:9:27) of special significance, in that they constitute two series of numbers in a 'proportion of porportions' —duplication and triplication.

Although Plato's *Timaeus*, in which this mathematical harmony was worked out, was well known to the Middle Ages it made little impact against the commonsensical rejection of it by the 'established' Aristotle. With the great reawakening of interest in Plato (a complex phenomenon to be dealt with in a later chapter), in the circle, first of Cosimo then of Lorenzo de' Medici, the acceptance of the idea became widespread by its incorporation in the design of Christian 'temples'. In his work *On Architecture* (written about 1450, printed Florence 1485) Alberti displays an intimate knowledge of the matter, and expresses the necessity for putting it into practice on the ground that it is indisputable that 'nature remains consistent with herself in every respect'. The acceptance of the ideal throughout Northern Italy is evidenced by the surprising appeal in 1534 to the Venetian philosopher, Francesco Giorgi (or Zorzi), to give his views on the disputed ratios of the dimensions of the church of San Francesca della Vigna, then being erected in Venice according to the model prepared by the distinguished architect, Jacopo Sansovino.

The foregoing architectural subtleties might have little reference to the problem we are here concerned with had not these mathematico-musical harmonies been associated in varying degrees with other potent ideologies—the Hermetic philosophy, the Great Chain of Being, and the hierarchical ordering of the cosmos by the late neoplatonic writer known as the *pseudo*-Dionysius. All these in varying degree reinforced the belief in a supreme cosmic harmony to which the microcosm, Man, must attune himself in all his undertakings, even though by some thinkers (p. 150) he was placed outside the *hierarchy*. None was in any sense a Renaissance 'novelty': belief in a pre-Christian non-Judaic body of cosmic wisdom emanating from the thrice-great Hermes ('trismegistos') is found in Lactantius; the hierarchy of elements and 'spirits' of *pseudo*-Dionysius was only an adaptation to Christian needs of the cosmology of the Neoplatonists; and the Great Chain of Being—though not the term itself—was tacitly accepted during the Middle Ages: its virtue for some thinkers being its origin in the natural philosophy of the more pragmatic Aristotle, who had exemplified the universal idea in the particular Scale of Nature applicable to all living organisms. Nevertheless, though not new, they came to play an increasingly important part in Renaissance thought.

Such, then, was the climate of opinion in which such diverse figures as Alberti, Leonardo da Vinci, Albrecht Dürer, Paracelsus and Girolamo Cardano could in effect, if not in so many words, claim that *Men can do all things if they will*: a claim to be echoed and codified at the end of the Renaissance by Francis Bacon and publicised in the Charter granted to the Royal Society by Charles II in 1662. The background has been sketched at some length as a warning against any attempt to answer what to the author appear to be the sterile questions as to 'the' origin of the new scientific spirit—in the groves of Academe or the courts of princes, in philosophical speculation or technological needs, among 'Platonists' or 'Aristotelians'. To frame the question in such terms is to assume such sharp distinctions as had no existence in fact. Like W. S. Gilbert's House of Peers the men of the Renaissance did nothing in particular but they did many things supremely well. But before their Promethean urge towards a more fruitful human achievement could bring about the 'revolution' of the seventeenth century it was to receive a further impetus from the powerful but muddied stream of 'natural magic', already beginning to flow in the *quattrocento* but to reach flood-level later. Its consideration will demand a separate chapter. To become effective the vision had also to become *organised;* the earliest instrument of this organisation was the introduction of printing from movable type, to be considered in the next chapter.

APPENDIX ON PERSPECTIVE

Owing to its origin among a closely knit group of creative artists there are few documents on which to base a chronological account of the Renaissance invention of linear perspective. Despite intensive study of the problem there is no general agreement as to the course of events. The justification for regarding Brunelleschi as having given the first concrete demonstration of the invention occurs in a work written by Antonio Manetti some years after Brunelleschi's death. The device was roughly as follows: A painting of the Florentine baptistry, as viewed from the duomo, was placed opposite a mirror in which, by looking through an aperture in the panel of the painting, an observer could see a reflection of the painting. The dimensions were such that the eye of the observer was placed in the same relation to the mirror as the eye of the painter (Brunelleschi) had been to the building, which thus appeared in full three-dimensional reality. There is strong presumptive evidence that this enterprise was prompted by the perspectival problems that faced Brunelleschi when seeking the most harmonious proportions for the churches of San Lorenzo and Santo Spirito—both fortunately still extant.

The subsequent history is confused by the attribution to Alberti of a

similar device set up within a dark box (hence the later confusion with
the *camera obscura* in which an actual image of a scene is projected on to
a screen) without reference to Brunelleschi. Probably the two 'demonstrations' were independent. What is beyond doubt is that Alberti's
critical analysis and mathematical demonstrations went far beyond
anything that Brunelleschi, even with Toscanelli's assistance, is likely to
have achieved. To the notions of orthogonal and parallel projection,
familiar to the Ancients from observation of shadows, etc., and applied
in the explanation of eclipses, was now joined that of *central* projection
by rays emerging from a point. The corollary, probably due to Alberti,
is the use in painting of the 'intersection plane' between eye and object,
on which relative disposition of parts can be sketched or measured.
Some have seen in this an 'anticipation' of coordinate geometry. If there
is any justice in this view, which is more than doubtful, it must be stressed
that it was no more so than the similar use of 'coordinates' in 'astro-
cartography' at Klosterneuburg about the same time (p. 31). There is in
fact a world of difference between the use of 'static' coordinates and the
expression of the properties of a curve by means of the relation between
the coordinates enshrined in an algebraical equation. Oddly enough
neither Descartes, nor his contemporaries, were able to achieve this final
step. The reader will search *La Géometrie* in vain for coordinates as
explicit as those employed by Marinos, the forbear of Ptolemy; Apollonios
came nearer to the modern use of coordinates than did Descartes.
Since the 'vanishing point' at which lines parallel in the object meet in
the picture plays a dominant part in their theory, what Alberti and his
associates did more nearly approach was *projective* geometry, first clearly
envisaged by Girard Désargues (1639) but systematically grounded only
much later by Gaspard Monges and Jean-Victor Poncelet.

4

'THE GUTENBERG GALAXY'

As with all the inventions that have increased the gap between Man and the rest of creation—speech, writing, fire, the wheel, the magnetic compass and gunpowder—nothing is known for certain of the inventor of printing or of the time, place, and circumstances of his invention. This is not solely due to the relatively primitive state of contemporary society, even for those inventions of major importance for our civilisation. It is only by virtue of retrospective study made possible by the greatly expanded world of scholarship that a similar fate has not attended the major discoveries of the last hundred years. Sometimes this has been due, as in the case of Max Planck's discovery of the Quantum Theory and Ernest Rutherford's 'splitting of the atom', to the fact that the discovery was too startling to be taken really seriously; sometimes, as when the President of the Linnaean Society reported nothing unusual during the year that saw the presentation of the papers of Darwin and Wallace to the Society, it had not seemed startling enough.

The introduction of movable type resembles the enunciation of the theory of natural selection. Printing, that is, the multiplication of copies from a single block, had been known for a century in Europe, a millennium in the East; speculations on the 'evolution' of living organisms go back at least to the Ionian natural philosophers; in each case a new 'paradigm' was introduced so fundamental as to change catastrophically Man's relation to his environment. Here the similarity ceases; for whereas printing appeared to owe nothing to 'science', Darwin's triumph rested upon the application of a novel hypothesis to a huge mass of correlated observations of organisms both in the state of nature and under the artificial control of domestication.

With the appearance of Professor Marshall McLuhan's epoch-making book, *The Gutenberg Galaxy*, it became necessary to reconsider the art of printing in relation to the human imagination; to ask ourselves, indeed, whether the usual assumption that print owed *nothing* to science isn't in effect to restrict the term 'science' to that aspect of human enterprise that took definitive shape only in the seventeenth century and whose character may very well have been to a considerable extent determined by the transformation that printing had made possible. If this were so, it looks as if the traditional assessment might have been caught in toils of circularity such as formerly enmeshed arguments about the origin and effects of the Industrial Revolution. It is fortunately unnecessary to admit all McLuhan's arguments (much less his perhaps rather extravagant claims concerning the 'electronic media') in order to accept his demonstration of the profound *theoretical* issues involved in the change from monobloc to transposable type. For whereas the former introduces none but *material* novelty (wood and ink in a new relationship) and a new *use* for the ancient winepress, the latter involves a tacit recognition of the extension of the potential inherent in the *visual* alphabet.

The extension involves not only the possibility of the synthesis of 'molecules' of meaning by the *linear* combination of meaningless material 'atoms' of various kinds but also the possibility of generating a large number of each kind of 'atom' from a single matrix. The 'molecular' potency of the alphabet was well understood by Aristotle, as was the further possibility of constructing infinite varieties of meaning by the manipulation of the 'molecules' (words) according to conventional rules (grammar). It has in fact been surmised for some time that the purely symbolical alphabet, combined with the grammatical conventions derived from Greek through Latin, has imposed upon Europe its characteristic modes of thought —modes that were unlikely to arise among races using a semitic language and impossible in the pictographic signs of early Egypt or the ideographs of China.

When it occurred to the polisher of mirrors, Iohann Gutenberg (if it was he), to 'realise' the atomic units of the alphabet in metal, he can hardly have suspected that this new technique would strengthen the grip on Western culture of the 'linear' mode of thought that during the age of manuscript and declamation had lain relatively lightly upon it. It is more than doubtful whether this fixation would have spread beyond a small circle of 'eccentrics' had not—significantly—the goldsmith, Johann Fust, distrained upon Gutenberg's type and press in payment of a debt, and, with his partner Peter Schoeffer, armed this interesting *idea* with the power of the rising technology. This speculation is of more than passing interest since such influence as there undoubtedly was worked both ways: the rapid expansion of the new enterprise (over 30,000 *works* in print by 1500) depended on facilities of metal-casting, some division of labour in well organised workshops, and of course capital; but also the nature of the task involved the creation of a new type of technology in which *entrepreneurs*, scholars, and manual craftsmen shared in the creation of the finished product. The framework of such a technology was already present in that characteristic Renaissance activity, the commercial copying of manuscripts. Though the great libraries, such as Urbino, were built up largely by 'house copyists' (*scrittori*) maintained by the prince, there was one man, Vespasiano da Bisticci, whose ability to carry out a 'crash commission'—on one occasion 200 works in twenty-two months—testifies to a high degree of organisation. He is said to have employed forty-five copyists, and his intimate dealings as a consultant with the majority of the most famous men of his age enabled him to compile a well-known collection of character-sketches.

The two-way flow we have already (p. 27) noted receives a further exemplification in the 'typographical explosion'. Though the first *effective* introduction of the *idea* took place in the German centres of metal-craftsmanship, it was in Italy, pre-eminently Venice, that the rapid increase of speed and scale of production began. The picture is somewhat complicated by the sack of Mainz in 1462, with the consequent dispersal of printers to other German cities and the displacement of Mainz by Frankfurt-am-Main as the main centre of German printing; but it remains true that nearly all the early presses in other countries were set up with the advice of Germans. It is also significant that viability of these presses depended on a complex of factors. The relative failure at Nürnberg, despite the city's pre-eminence in metalwork backed by a cultured governing class, may have been due to the absence of any institutional scholarship such as a university could provide. But that a university was not sufficient is shown by the slight and ephemeral early activity at Oxford and the late and almost equally ephemeral press at Cambridge despite the humanistic surge there before the end of the fifteenth century. The case of Venice was unique (it usually was!): though lacking a university within its island confines it was within easy distance of Padua and Vicenza among its own landward possessions and of the vigorous young university of Ferrara, reaching its greatest celebrity just at the critical time for the establishment of printing.

That the introduction of printing could have had no influence on the Italian origin of the Renaissance needs no emphasis. That its influence on the 'New Learning' up to about 1500 was almost negligible is shown by the fact that in the great libraries, such as Urbino, no printed work would have got further than the *quattro-*

cento equivalent of the coffee-table; but it is shown more specifically in most subjects by the absence among printed works of more than a handful elaborating new ideas. In mathematics, apart from the beautiful *princeps* of Euclid's *Elements* (Ratdolt, Venice 1482) and a few 'arithmetics' for the counting house, there was nothing before Pacioli's *Summa* (p. 49); the most original work of all written during that period, Regiomontanus's *Of all sorts of Triangles* (1464), was not printed until 1533. A similar, though not quite so black, picture appears for astronomy: the only 'modern' work, Peurbach's *New Theories of the Planets*, was indeed available in two editions, but the indispensable *Almagest* of Claudius Ptolemy became available only in 1528 (Latin) and 1538 (original Greek); even the medieval translation from the Arabic version appeared only in 1515. The forward-looking works of Nicholas of Cues (also of great theological significance) were not printed until 1488, and then inadequately; those of Archimedes not until much later. Most significant of all perhaps is the fact that the five most *scientific* works written in the Middle Ages—Pierre de Maricourt's treatise on the magnet; the Emperor Frederick II's wonderful work on falconry (conceived in a systematic and critical manner wholly foreign to medieval writers on the living world); Leonard of Pisa's no less remarkable studies of the 'new' Hindu-Arabic arithmetic and algebra, including his own still famous Fibonacci series; and the advanced works on optics by Witelo and Dietrich of Freiburg—were printed very late, or, as far as the Renaissance is concerned, not at all. By contrast, the fate of another medieval work, *On the Sphere of the World* by Ioannes de Sacrobosco (John of Holywood), that achieved *thirty* editions before 1500, points to the reason why. For, whereas the five neglected works would have appealed to a rather limited clientele, the latter had been required reading in most universities for a couple of centuries; and universities then were even less amenable to change than they are now.

Printing was indeed a form of trade. The early printers, who were also the publishers, though mostly men of scholarly attainments and ideals (Erasmus found no more congenial home than that of the elder Froben at Basel) were forced to make a living by selling their wares. Authors might rely on patronage from princes; but, until the immense propagandist power of the new invention had been demonstrated by the protestant reformers, printers could look for no support from institutions, whether secular or ecclesiatical; they had no alternative but to exploit existing demand. Only a super-ficial glance through a representative collection of 'incunables' (books printed before 1501) is necessary to reveal what that demand was, namely, works of the Fathers and great 'clerks' (theology was by far the greatest component of even the Urbino library), breviaries and missals, the Latin classics—especially Cicero and Virgil—and, as Regiomontanus demonstrated (p. 33), calendars including data for astrologers. Though through the enterprise of that 'good European', William Caxton, printing was introduced into England little more than twenty years after its German beginnings, neither in his own editions nor in those of his partner and successor, Wynkyn de Woorde, were there any except such as would appeal to a clientele of piety and taste but without pretensions to solid learning. The few customers who set their intellectual sights rather higher would in any case be in close touch with continental sources.

One 'science', apart from those subsidiary to astrology, that had commercial possibilities was geography, whose 'bible' was Ptolemy's *Cosmography* (also known as *Geography*). This not only appeared (though without the highly important maps) at Vicenza as early as 1475 but achieved four other *incunable* editions in Italy and two in Germany (Ulm), all with maps, the later ones incorporating new knowledge.

To what has just been said the printing of works on Medicine appears to be a striking exception. Among the medical works printed before 1501 about 125 authors are represented. Of these forty-seven had been active before the fifteenth century. The remaining seventy-eight might thus be regarded as 'contemporary' though not necessarily still alive when their works were printed. This proportion of 'contemporary' to 'traditional' authors whose works were selected for printing was far greater than in any other 'scientific' subject. But the circumstances relating to the study and practice of Medicine were peculiar in several respects. The 'Higher' Faculty of Medicine was the only way to advanced study of natural knowledge beyond the traditional (Aristotelian) 'physics' of the Faculty of Arts, in which most of the traditional mathematical *quadrivium* was by then playing a much smaller part; the stimulus towards the study of astronomy came mainly from the medical schools. It is not surprising then to find more creative activity in the medical subjects. But this was not all; Europe was periodically beset by visitations of Asiatic plague, and the last decade of the *quattrocento* saw the rapid spread of the *morbus gallicus* or 'Great Pox'. Over thirty of the authors reckoned as 'contemporary' were so by reason of their tracts relating to these menaces—clearly a field in which the new means of rapid dissemination could play a decisive part. Nevertheless, the 'progressive' effect of printing, even in this context should not be exaggerated. A book, none of whose contents (with the notable exception of a plague tract) was less than a century old, was found worthy of being printed in various forms and languages eight times in ten years. Although one of these versions was in Italian and three in Spanish, the remainder were in Latin, which disposes of the possibility that it was merely a popular line of the sort later to be known as the 'Family Physician'.

In the foregoing hasty glance over the first half century of printing an attempt has been made (without, it is hoped, a biassed selection from the great mass of evidence) to show that despite its rapid expansion the direct impact of printing on the 'New Learning' was almost negligible. The beginnings of the coming revolution in natural knowledge were however already established, though they were first made apparent in humanistic scholarship. Of these, strict reproducibility was perhaps even more fundamental than speed and cheapness of reproduction as such. As long as speech and manuscript were the only means of diffusion of knowledge it was virtually impossible to produce a 'definitive text' of a great work. Even apart from the appalling hazards of translation of those ancient classics that had been transmitted through the medium of Arabic the mere process of copying provided temptations unlikely to be always resisted. Omissions were bad enough, but doubtful passages would be 'interpreted' by the copyist with a consequent risk of distortion of the original meaning. By the time successive copies of a 'text' had been made the sense might have undergone such changes as to render it unrecognisable to the original author. Part cause, part consequence, of this mode of diffusion was that in the medieval method of teaching by lecture and disputation the distinction between the advancement and the transmission of knowledge was apt to become blurred. Such a state of affairs greatly favoured the latter. But, on the other hand, we should not fall into the error of taking the term 'commentary' in its literal modern sense: much of the most important critical and even innovatory medieval thought was handed down in the form of a 'commentary' on (e.g.) Aristotle's *Physics*.

To decide in what field of knowledge there first arose serious dissatisfaction with the uncertainty as to what the 'authorities' really had written is too big a question to discuss here; there can however be no question but that the remedy was first taken in hand by the humanists, men who gave *priority* to linguistic precision and terminological exactitude. It is necessary to stress *priority* to remove the too common implication that mere elegance and grammar counted for more than subject-matter.

It is perhaps not unreasonable to suggest that this widespread recognition of the necessity for the multiplication of copies of a single definitive text was the principal factor in making printing viable throughout the world of learning. It is unnecessary to emphasise that though in its first fifty years printing had little effect on the dissemination of scientific ideas, once established it had an influence it would be difficult to exaggerate. Henceforth *progress* could be based on a position clearly taken up and rapidly criticised by contemporaries in the same field, but widely spread in regard to centres of learning. Reference to a bibliography of sixteenth-century scientific books will reveal how quickly 'revised and corrected' editions followed an original that had caught the professional eye.

A side-effect of this possibility of producing at relatively low cost any number of identical copies was the 'text-book'. By 1565 Leonhart Fuchs, Professor of Medicine at Tübingen, had lost count of the number of editions of his textbook, *Five 'books' of the Institutes of Medicine*, but he thought it was the sixth to which he was writing a dedication. His doubts, and the complaints of his readers to whom he was replying, were probably due to the fact that during the previous thirty-four years he had written a succession of books on more or less the same subject under different titles, the later ones being revised and considerably enlarged versions of the earlier. Matters were further complicated by the fact that in the absence of any international copyright, other than the 'privileges' granted for a single book by princes and ecclesiastical potentates (usually the pope), editions or variants appeared from numerous presses in different parts of Europe. In this respect the immense expansion of printing was not an unmixed blessing, some of the advantages being thereby nullified.

Even before the demand for medical textbooks had been exploited by the printers a movement towards the 'packaging' of knowledge—so significant for the commercial viability of printing—had been discernible in the growing emphasis in the Faculties of Arts on *logic*. So far from admitting the necessity for intellectual maturity for its comprehension the universities tended more and more to regard logic as a propaedeutic to all other knowledge (except grammar). The beginnings of this policy may be traced back to the *Summulae logicales*, written about 1250 by Peter of Spain, which became the almost universal 'primer' of logic until the late *quattrocento*, when the German humanist Rudolph Agricola replaced it by a work based more closely on the Greek text of Aristotle, though still on that part of it known as the *topics* rather than on the more 'scientific' *Prior Analytics*—a restriction made necessary by the extreme youth of the victims of the system. It is still not widely recognised that the strict—and in some ways far more 'modern'—canons of the latter work were never seriously applied in the so-called 'Aristotelian' logic of the Middle Ages.

This standardisation of exposition presented a great opportunity to the printers. It has indeed been argued by Fr W. J. Ong that it may have been an important factor in the precipitation of the invention, for which all the material conditions had long been available. The ready response and technical competence of the more enterprising printers in turn made possible the more extensive

development of what Fr Ong has picturesquely called the 'Peda-
gogical Juggernaut', the ultimate benefit of whose triumph has
latterly been called in question. Since this raising to a new dimension
of the relation between science, print-technology, and society was
largely due to the influence of one man, Pierre de la Ramée, within
the rather special circumstances of Renaissance France, its further
consideration will be deferred to Chapter 8 which deals with that
society.

5

SCIENCE AND THE NEW WORLD

When Christopher Columbus addressed a letter to the Director of the Bank of St George at Genoa in 1502—almost ten years after his landfall in the Bahamas—he signed himself 'Great Admiral of the Ocean Sea and Viceroy and Governor of the Islands and Mainland of Asia and the Indies belonging to the King and Queen my sovereigns . . .'. Four years later he died an embittered and frustrated man. After four attempts he had failed to see the golden palaces of Cipangu (Japan) or to present his ambassadorial credentials from Ferdinand of Aragon and Isabella of Castile and Leon to the Great Khan (who had long ceased to function) on the Asiatic mainland. On the contrary, by his nagging persistence in a literal interpretation of the terms of the original royal charge he had so smirched his image at court that the only official record of his death is in the minute attached to correspondence relating to his claims. The verdict of history has been different: his actual achievement has been rated as incomparably greater than the demonstration of an alternative route to the (East) Indies, at least as great, and probably more significant, than the recent 'moonfall'. And in any case the sea-route to India had been achieved by Vasco da Gama in the employ of the rival Portuguese crown before Columbus had (unknown to himself) struck the South American mainland.

In the account that follows no attempt will be made to present a chronological sequence of what J. H. Parry has admirably called the 'European Reconnaissance', but only to try to assess its relation to the contemporary science and the further implications for Renaissance society.

The original royal charge to Columbus was to 'discover and conquer certain [otherwise unspecified] islands and mainland'. The

C

belief in the possibility of discovering (or rediscovering) islands in the Atlantic was a well-established tradition: the fabulous 'Antillia' was hopefully marked on some maps; only a few years previously the far westward Cape Verde Islands had been positively sighted by the Italian Alvise Ca' da Mosto in the employment of the rival Portuguese. But the 'mainland' referred to in the charge could have been no other than the eastern seaboard of Asia with its offshore island of Cipangu. The discovery of the 'New World' must then be regarded as wholly unsuspected—a very lucky accident. What then can such an outcome have to do with science? The fact that among the extant books from the library of Columbus four were of the best available printed works relevant to the enterprise and that they were liberally spattered with notes in his hand would be sufficient evidence that the scientific factor cannot be neglected. An accurate assessment of the picture is indeed impossible: not only are too many pieces missing but the authenticity of some of those available is still open to question. On the acceptance or rejection of these dubious pieces depends the sort of picture—and there have been many—that can be constructed of the relation of the enterprise to the 'science' of the day.

Among these contested documents the most important are two letters from the foremost mathematical cosmographer of the day 'Paulus physicus' (Paolo Toscanelli, see p. 29). One of these (addressed to Fernao Martins, Canon of Lisbon cathedral) gave reasons for his belief that the coast of Portugal was no great distance from 'Cipangu' (Japan). Alleged to have been accompanied by a map, this letter exists in what is believed to be a contemporary copy. The second letter exists only in the form of an *alleged* copy in a book by Columbus's son, Ferdinand, and also in a later work by Bartholomé de Las Casas, son of a shipmate of Columbus. In this 'letter' Toscanelli 'replies' to a query by Columbus and encourages him in his projected voyage. Columbus had been dead many years before either of these books was written, and there is no evidence that he ever referred to this alleged letter. Since Martins was present at Todi when Toscanelli witnessed the will of the dying Nicholas of Cues and almost certainly met Toscanelli in Florence the balance of opinion favours the authenticity of the letter Martins is supposed to have received, probably at the instigation of Alfonso V of Portugal, but rejects the personal letter to Columbus as a fabrication of Ferdinand to bolster the prestige of his father as a 'scientist'. Whether or not Columbus *saw* the letter to Martins can never be *proved*; but it seems not unlikely.

The significance of the 'genuine' letter of Toscanelli lies in his use of the testimony of the great merchant-discoverer, Marco Polo (d. 1324), to 'correct' Ptolemy's estimate—180°—for the eastern extension of Asia. This bold challenge by a humanist, whom Pico della Mirandola (p. 149) referred to as 'certainly learned in medicine, but principally in mathematics, Greek and Latin', to the reliability of a classical authority, and especially his appeal to the report of a merchant, whose actual experience of the Far East was by then commonly ignored or rejected, provides us with a further reason for regarding Toscanelli as a focus of *quattrocento* activity in respect of natural knowledge. If Columbus really did know of Toscanelli's letter to Martins it is most likely that he got a sight of it when in Portugal, where in 1484-5 he was trying to get the support of the crown. His claim to such support could have been based on the fact that for many years he had been earning his living in partnership with his brother, Bartholomew, as a chart-maker in Lisbon, or sailing under the Portuguese flag. His 'correspondence' with Toscanelli, if it ever existed, must in any case have been completed before 1482 when the latter died.

As far as the enterprise of Columbus himself is concerned any knowledge he may or may not have had of Toscanelli's views is of importance only as 'learned' confirmation of his own 'hunches'. For he actually possessed the works (in Latin translation) of Marco Polo (printed in 1484), and in these he would have discovered that the author not only added about 30° to the eastern extension of the Asiatic mainland but he knew, as Ptolemy did not, of the wonderful island of 'Cipangu' that he believed to be about 1500 miles out to sea. All this would have been to the dedicated Columbus a further 'improvement' on the estimate of Cardinal Pierre d'Ailly, whose *Image of the World* (based on the thirteenth-century work of Roger Bacon) was heavily glossed by Columbus, and who in an accompanying tract had also dared to criticise Ptolemy for having *under*estimated the extent of the mainland.

All these estimates—even Ptolemy's—were hopelessly *over* the mark; but Columbus was so very sure of his case that when, as in the near contemporary *Universal History* of Eneo Silvio Piccolomini (later Pius II) he found more conservative estimates, he simply ignored them.

So much then for the 'scientific' basis of the 'Enterprise of the Indies'. In fact this is only the *cartographical* fiction; there was still the geodetic problem of the *length* of the 'degree'; for though his course might be set in degrees his track would be measured in nautical miles. Here the Muslim astronomer Al-Fagarni comes on the scene and is pressed into the service of Columbus by the absurd assumption that the Arab had used the short Roman mile, thereby reducing the arc of a degree from the actual sixty nautical miles to forty-five. By means of these successive 'weightings' the transit from the Fortunate Isles (Canaries) to the Indies had been made to appear as no great matter, since Japan could be counted on as a certain 'staging post' if (as suggested by Toscanelli) the apocryphal island of Antillia had not already served for this purpose!

If Columbus's estimate of the necessary 'westing' to reach the Indies was based on scientific fiction the rejection of his plan by the Portuguese experts may well have been founded on 'scientific ignorance'—a cynical disbelief in the very existence of 'Cipangu' or even of 'Quinsay' (Hangchow) which had no 'scientific' warrant (that is, they were unknown to Ptolemy) but only that of a 'traveller's (Marco Polo) tale'. If this was so, it is an example of the power of ancient 'authority' when set against that of the systematic and mainly sober account of an eyewitness. But it is also an example of a correct decision taken for the wrong reasons: Columbus estimated the distance from the Canaries to Cipangu to be about 2400 miles; measured along the 28° parallel it is actually more than four times this amount.

From this brief summary of the 'theory' by which Columbus sought to justify his enterprise he emerges as a fanatical visionary in whom faith was made to triumph over reason; but he was also an experienced seaman who had profited from the discoveries of the much more systematic Portuguese voyages. The glosses in Pierre d'Ailly's *Image of the World* (Columbus's cosmographical bible) testify to his frequent navigational observations in Portuguese service down the west coast of Africa; most significantly (and inconsistently) he rejects Pierre's repetition of the universal medieval belief (based on classical 'reasoning') in the uninhabitability of the Torrid Zone with the words '. . . it is even very populous, and under the equator is the castle of Mina of the most serene king of Portugal which I have seen'. The ruins of the castle are still to be seen on Cape Coast in Ghana—actually about 5°N. Before 1467 the Guinea Coast was as much a 'New World', never before known and on the 'authority' of ancient 'science' unknowable, as was the continent of America; twenty-six years later Columbus was in Lisbon when Bartholomew Diaz returned from his voyage beyond the Cape of Good Hope thus shattering another ancient scientific myth—that of a landlocked Indian Ocean.

Of comparable importance in support of his plan was the occasional appearance, well known to seamen, of driftwood on the shores of the Azores; Columbus himself had seen canes far thicker than any known on the African coast; the 'sea-beans' regularly cast up on the shore are still known in Porto Santo as *favas de Colon*; and in 1869, after a storm of exceptional severity, trunks of the Central American Cupo tree were seen. Of course, there was no guarantee that these objects came from the 'Indies'. The Cape Verde Islands—about five hundred miles west of the nearest land

(Dakar)—had been sighted by Ca' da Mosto when Columbus was only five years old; there might well be other *islands* beyond these. But with his usual inconsistency Columbus was ready to find in Ptolemy, Aristotle, or even Seneca hints of the fauna and flora likely to be found in the Indies, and then, as well as during his subsequent discoveries, *see* the resemblances he was looking for. In fact he showed himself to be an excellent observer of bird and beast and vegetation, careful at all times to give a clear description before drawing a frequently fantastical inference as to its significance.

It was this skill, and the habit of systematic recording of his daily experiences rather than any technical knowledge of cosmography, that enabled him to justify the trust he ultimately won from the Spanish sovereigns—Queen Isabella in particular—to 'discover and conquer certain islands and mainland'. The story of his first voyage and landfall is well known. Actually he could hardly have had an easier voyage—as such it was not to be compared, either in extent or difficulty, with Vasco da Gama's circuit of Africa and transit of the Indian Ocean with no sight of land for three months and twenty-three days respectively. The return voyage of Columbus was a very different matter: on this, as the checking of every entry in his log made possible by the Harvard Columbus Expedition of 1939–40 revealed, he displayed uncanny powers of navigation by dead reckoning and, in the terrible hurricane that overtook the fleet, the superb seamanship that never deserted him in the tricky coastal and inter-island navigation of the later voyages. Nevertheless it is doubtful whether the caravels would have survived the return voyage but for a gross error of course-setting that took them into the region of the prevailing westerlies, of which no one then could have had any knowledge.

Though the actual discovery of America involved no scientific knowledge beyond that of any reasonably well-educated contemporary, it must not be supposed that Columbus was wholly unaware of the considerable advances in the *science* of navigation that had made possible the epoch-making discoveries of the Portuguese; in fact his travelling kit included a sand-clock, magnetic compass, marine astrolabe, simple (not 'reflecting') quadrant, and the *Ephemerides* of Regiomontanus (p. 37). He was fully aware of the fact that not only does the magnetic compass commonly point east or west of true north, but also that this variation (better called 'declination') varies in a fairly regular manner in relation to different points on the Earth's surface. Since the change is only gradual, making no significant difference in observations made in the coastal navigation of Western Europe, some compasses had a 'built-in' correction that might be a source of error if not allowed for on a long voyage. The astrolabe and quadrant were primarily for 'fixing' latitude as a check on dead-reckoning. This was done by finding the elevation of the Pole Star above the horizon—a simple operation on *terra firma* but impossible in a rough sea. There is no record that Columbus ever used his astrolabe, and his observations with the quadrant even on land were several times seriously in error owing to his mistaking another star for Polaris in the unfamiliar sky of the more southerly latitude of the 'Indies'. There was not then, nor for another two centuries, any straightforward method of finding the longitude at sea except on the occasion of an eclipse. It was for this reason that Columbus carried a copy of Regiomontanus' *Ephemerides*, and on one occasion during his last voyage used it to good effect to scare the natives! This prediction of a solar eclipse was probably the nearest Columbus ever came to justifying his own claim to be well versed in celestial navigation.

In the discovery of the 'Indies' (as distinct from his later reconaissance of the northern shores of South America) the problem of the *disappearance* of Polaris never arose; for the Portuguese voyages to the Guinea Coast and beyond, on the other hand, this became a major problem. Important as had been the missionary fervour of his great uncle, Prince Henry 'the Navigator', it was the more sophisticated John II who established the mathematical institute by which alone this problem could be tackled. The unique property of Polaris is its relative brightness and easy recognition by prolonging the line of the 'Guards' of the unmistakable 'Plough'. Admittedly Polaris does not coincide with the true North Pole of the heavens but its relative proximity thereto allows of the application of a simple correction according to its position relative to the circling 'Guards'; this was known as the *Regimento del Nord*. Unfortunately, when Polaris approached invisibility as the navigator neared the equator, no 'southern' star appeared to mark even approximately the South Pole; the alternative—the Sun—would fill the bill admirably but for the fact that it does not cross the heavenly vault at the same rate as the stars, and worse, its apparent movement through the stars is not uniform. So the *Regimento del Sol*, based on systematic observations from a fixed point, had to be introduced—a procedure far too difficult for the average sea captain to apply. Of the earliest introduction of such a *Regimento*, or of the tables on which it was based, nothing can be said with certainty, since the closest 'security' precautions were taken in respect of all Portuguese marine enterprise. Solar tables prepared by Regiomontanus were available in print by 1495; but since these were primarily for astrologers it is unlikely that even if MS copies had been available earlier they would have been of much use to navigators. The earliest *printed* '*Regimento*' can be referred to about 1509; but it is hardly conceivable that the long 'legs', out of sight of land, traversed by even

Diaz, would have been undertaken without some means of fixing latitude south of the equator. No Portuguese documents relating to navigation were published until long after their composition; but recent Portuguese scholars, albeit with occasional extravagant claims in support of vanished national glories, have provided convincing evidence of the availability of the *Almanach perpetuum* (printed 1496) of Abraham Zacuto, professor of mathematics at Salamanca, whose pupil, José Vizinho, was a leading member of John's council of experts.

Such then was the new *science* of navigation that, coupled with the striking advance in ship design of which the caravel was the fruit, the superb art and unswerving faith of Columbus, and a great deal of luck had within half a century made 'one world' of what previously had been four almost isolated fragments. To such a unity the term '*New* World' might be most aptly applied. When Columbus on his third voyage told his sovereigns of the *otro mondo* ('other world') ready for their conquest, he was still convinced that the newly discovered territory was an extension of 'India', though whether continental or a huge island he was at that time uncertain; a similar interpretation may be put on the use of *mundus novus* by early chroniclers of his discoveries. The identification of 'the New World' with America became customary only some years after the cartographer, Martin Waldseemüller, had shown in his world map of 1507 a continent, in its north-south extension comparable with Europe and Asia, and suggested that the southern part be named after Amerigo Vespucci, its alleged discoverer.

With the still disputed claim that Vespucci was the 'real' discoverer of America we need not concern ourselves. Based at first on spurious printed works, in which letters of Vespucci were garbled in such a manner as to imply that he was convinced of the discovery of a new *continental* mass before Columbus had suspected its great extent, this claim had a popular appeal, blown up by successive editions made available by the new 'mass medium'. This has served to cloud a number of significant historical facts. Vespucci was a highly educated business agent in Seville in the service of the Medici bankers, and, though no professional seaman, he knew a great deal more about cosmography and the theory of position-finding than any but the most exceptional sea captains: in short, one of the first of a new race of men whose importance, especially in Elizabethan England, must not be overlooked. In two letters discovered in Florence in the eighteenth century and addressed to Lorenzo di Pier Francesco de' Medici, written from Seville in 1500 and from Lisbon in 1502 respectively, he gave an account of his observations of fauna, flora, and native customs, which are a model of scientific objectivity. Knowing Lorenzo's informed interest in cosmography, he set forth data on the solar shadow and the polar stars that proved that even in the earlier of the two voyages the ship he was travelling in sailed for several days in the Torrid Zone some degrees south of the equator, which 'confuted the opinion of the majority of the philosophers, who assert that no one can live in the Torrid Zone. The health of the crew and the well-being and large numbers of natives proved the contrary . . . let it be said in a whisper experience is certainly worth more than theory'. By comparing observations on the lunar conjunctions of planets with the calculated times in the 'almanach of Giovanni da Montereggio' he estimated his longitude: the attempt is of more interest than the result, which was, not surprisingly, seriously in error. In the letter relating to the second voyage he claimed to have reached 50° South—a greater distance south of the equator than Seville is north of it. This convinced him that (South) America must be a continent having no connection with the 'Indies'. This was confirmed eleven years later when Vasco Nuñez de Balboa (not 'stout Cortez'!) 'with eagle eyes gazed at the Pacific'. The enormous extent of this new continent was finally established when, after Magellan had passed the southern strait, Giovanni Verrazano failed to find a 'sea passage to Cathay' between the Carolina coast and Newfoundland, concluding that the latitudinal extent of America is greater than that of Europe and Africa between the North Cape and the Cape of Good Hope. Magellan was killed in the Philippines, but Sebastiano del Cano fulfilled Columbus's dream of reaching the 'Indies' by the Western Ocean and thus became the first captain to sail round the world.

If the discovery of a New World depended more on faith, art, and resolution almost to the limits of physical and moral endurance, than on science in the modern acceptation of the term, the consequences of the discovery have been less thoroughly studied and are more difficult to assess. This assessment must in any case be attempted in the light of the aims of the enterprises, explicit or implicit: sometimes delivered with a brutal frankness—'gold and spices'; sometimes less brutally but perhaps with an equal sincerity—'the conversion of souls for the enlargement of Christendom'; and less overtly, but at least by implication, the establishment of trade-routes to the known sources of silk and spices avoiding the Mediterranean Sea, commanded during the second half of the fifteenth century by the fleets of the Ottoman Turks. Of course the desire of the individual explorer for romantic adventure and personal fame— as was certainly the case with Columbus—or natural curiosity— probably the principal stimulus for Vespucci—must not be overlooked. Such examples are symptomatic of a change of attitude from that of an age that had 'forgotten' Marco Polo and showed itself grossly ignorant of the customs of the courts of the Great

Khan, which had in any case by then ceased to exist. Nevertheless each candidate for patronage had to offer hope of a tangible return for their money. This may well account for the statements by some scholars (based it would seem on a rather superficial reading of his journals) that Columbus showed only a perfunctory interest in the geographical features and natural history of the islands he discovered while giving accurate and fairly comprehensive accounts of their human inhabitants. The greater emphasis on human 'souls' (or, alas, slaves) was natural enough in the circumstances. Nevertheless, on the rather rare occasions when he did describe striking natural features he showed a remarkable objectivity and accuracy for one whose occupation when not at sea was cartography. He must have expected to find the monsters that were the creatures of the imagination of the authors of medieval bestiaries and the like; yet he saw no such animals or such men. Capable of interpreting his observations as evidence for the Garden of Eden and similar fantasies, he nevertheless reported only what he saw. An example cited by Morison is especially revealing: this was Columbus's account of a tree having 'branches of different kinds all on one trunk, and one twig is of one kind and another of another and so unlike that it is the greatest wonder in the world. . . .' Only a skilled botanist could have recognised that, as Morison suggests, 'the tree was full of the different parasites [epiphytes?] that are common in the West Indies'. An observation of great scientific significance was the absence of any mammals on the islands. As to failure to give *precise* accounts of geographical features it is sufficient to repeat Morison's assurance that many were good enough to enable the features to be recognised from the air.

The case of Vespucci, if we can trust the much more meagre information we have about him, is in some respects more significant. As commercial agent for the Medici he had no material incentive to go a-voyaging, first with the Spanish, later with the Portuguese fleets. His position was rather nearer to that of the young Joseph Banks who in 1768 sailed with Captain James Cook to the Antipodes at his own expense and to enlarge his knowledge of the world flora. Vespucci's hobby could best be described as cosmography, with the necessary concomitant astronomy; hence his terrestrial observations were much more limited except with respect to the customs of the natives; but his original method of fixing longitude by planetary conjunctions, though superseded in general practice, remained in text-books of navigation until at least the eighteenth century.

These two examples must suffice to show that there was a growing sense of awareness of the immensity of the range of natural objects and of an aesthetic pleasure to be gained in their investigation; and that when this was coupled with a more strictly empirical approach than Columbus adopted there was revealed a *basic* uniformity underlying a bewildering variety of detail. This would be a hazardous inference from so little evidence; but some degree of confirmation may be found in the converse relationship—the immense stimulus to natural enquiry provided by the huge access of unfamiliar material to exercise it upon—that became evident by the middle of the sixteenth century (p. 110). Nothing is more striking in the history of the study of animate nature than the admission by the Tübingen professor of Medicine, Leonhart Fuchs, that the study of the *Materia medica* could be advanced by walks in the open air of the alpine slopes carpeted with flowers to delight the eye.

A striking and little known piece of evidence for the widespread growth of a more empirical mode of thought is revealed in a passage in a long letter written from Buda in Hungary in 1514 by Giovanni Manardi (p. 85) on the general medical topic of the possibility of life in the Torrid Zone. Having claimed that this had been put beyond dispute by the voyages of the Portuguese (*Lusitani*) to the extreme west of the Atlantic as well as in the East 'where neither the sea nor any other circumstances is a bar to human habitation', Manardi clinches the argument as follows: 'If anyone prefers the testimony of Aristotle and Averroes to that of men who have been there, there is no way of arguing with them other than that by which Aristotle himself disputed with those who denied that fire was really hot, namely, for such a one to navigate with astrolabe and abacus to seek out the matter for himself.' There are several points of interest in this passage. It is surprising to hear an 'academic' not only ready to refute Aristotle by reference to current experience but to point out that such is what Aristotle himself would have done: the mass of error perpetuated in the name of 'the Philosopher', as the medieval scholastics were wont to call him, was the work of lesser men; in this respect Aristotle was far from being an 'Aristotelian'. But Manardi was no ordinary 'academic'; when he wrote this letter he had been for many years removed from academic circles as private physician, first to Giovanni Pico della Mirandola (p. 149) and later to King Ladislao of Hungary. And the university he had graduated from, and to which he would finally return as professor of Medicine, was no ordinary university: Ferrara had indeed been given a 'new look' by Leonello d'Este as early as 1445 and was still so 'modern' in its attitudes in the early *cinquecento* that the violently *anti*-academic Paracelsus (p. 87) was proud to claim (on slightly dubious grounds) laureation there in Medicine and Surgery. Moreover the earliest known letter to Italy announcing the discoveries of Columbus was forwarded to Ercole d'Este, the reigning Duke, a little more than a month after the former's return, though it had already passed through other hands. On Ercole's prompt request for more detailed information his envoy in Milan obtained what was probably a copy of Columbus's first letter to the sovereigns. Of course, news ultimately reached other Italian courts direct, but it was at Ferrara that it created the most informed interest.

In view of the fact that the Welsers of Augsburg and other German merchant-bankers maintained agencies in Lisbon it is difficult to explain why the news of Columbus's discovery reached Nürnberg, the centre of learned interest in cosmography (p. 39), as much as three months after it was known in Italy, and why even then it seems to have aroused little interest. It is possible that the Germans' eyes were too narrowly focussed on the exploits of the Portuguese, for the Welsers had been among the first Augsburg merchants to be associated with the Portuguese fleets opening up the spice trade.

In assessing the speed and degree of appreciation of the profound implications of the discovery of the 'Indies' allowance must be made for the fact that then as now it is the relatively trivial titbits of 'human interest' that are likely to travel fastest and farthest. Among these, even compared with gold and spices, eyewitness reports of men, and especially women, going about their lawful occasions clad only in the scantiest *cache-sexe* must have aroused the greatest interest. This was twofold: not only would the information appeal to prurience (Morison reminds us that naked women were a much rarer sight in the sixteenth century than they are today) but it also confirmed the not uncommon belief that in the 'East' there remained some remnant of the 'Earthly Paradise'. The peaceful and trusting nature of the first found natives (naturally long since exterminated by their Christian discoverers) confirmed this expectation. The prospect of new foodstuffs and new drugs had also a striking news-value, heightened in respect of the latter by the early 'importation' of what appeared to be a new disease of appalling and widespread impact. Since this affected the development of Medicine in many ways other than the purely epidemiological it will find a more suitable context in Chapter 7.

6

SCIENCE AND POLITICAL THEORY

In Adam Smith's view 'the discovery of America and that of a passage to the East Indies by the Cape of Good Hope are two of the greatest and most important events recorded in the history of mankind'. But he was writing nearly three centuries after the events. To contemporary Europeans the realisation of the nature and extent of their importance was impossible: indeed the recent claim of the Mexican historian, Edmundo O'Gorman, that America was not discovered but rather invented by Europeans expresses an insight beyond mere paradox.

In the previous chapter some examples of the more evident responses to the discovery of the New World have been adduced. In addition to those rather confused reactions—in part foreshadowing the empirio-critical attitude characteristic of modern science, in part reviving romantic longings for a Golden Age behind the arid rationalism of the 'barbarian' schoolmen—there remains to be discussed the even more complex and largely speculative question of the impact of the discovery on what Burckhardt called the 'Renaissance state as a work of art'.

Though the emergence of the modern *nation*-state can be discerned first in France and England (Chapter 8) rather than in Italy it was the Florentine, Niccolò Machiavelli, largely as a consequence of the impact of France on Italy, who first made explicit the conditions under which a 'state', in the sense of a community dominated by a centralised power, could be expected to succeed. With the evaluation of that literary masterpiece, *The Prince*, as a reasoned and effective guide for the 'principals' of that centralised power we are of course not concerned. Nevertheless, though it almost entirely ignores the outstanding 'scientific' novelty, gunpowder, which even as he wrote was revolutionising the art of war, a persuasive case has been put forward, notably by Leonardo Olschki, for regarding the structure of Machiavelli's argument as a close approximation to the method later to be adopted in the natural sciences. The basis of this claim may be summarised under the following heads: (1) an explicit aim to 'represent things as they are rather than as they are imagined, or, it might be added, as good men think they ought to be'; (2) the assumption that men's deeds take place in a world of *unvarying order*, making prediction possible; (3) the admission of the impossibility of total knowledge, having as its corollary the necessity for recognising that decision is a choice between uncertain outcomes; and (4) since circumstances are constantly changing, generalisation is impossible. These heads themselves exemplify the elements of uncertainty, since (2) is literally inconsistent with (3) and (4); but this very inconsistency at the formal level is, Olschki maintains, a measure of the necessity for an 'inductive' element, in which, while an ultimate order has to be assumed, the consequent risk of error in prediction must be narrowed by continual resort to the results of experience.

In the single-minded advocacy of the pioneer Olschki exaggerates the closeness of the parallel, just as Machiavelli in the role of pioneer of political science is guilty of some backslidings and, even more, inconsistencies in the application of his canons to the concrete case of Italy. What both writers omit is the fundamental importance in the natural sciences of *experiment* in addition to *experience*; or where, as in astronomy, experiment is impossible, the repetition of experiences under conditions as nearly as possible identical. Of course even the latter expedient is impossible in establishing a theory of politics; and it is greatly to Machiavelli's credit that he recognised this and the limits it set upon effective generalisation. There remains a further feature of Machiavelli's thought that Olschki rightly considered at some length: this is the part played by *Fortuna* in human affairs.

The value of the penultimate chapter of *The Prince* lies in an analysis of the element of Fortune rather than in the conclusiveness of the argument. 'Because free choice cannot be ruled out' Machiavelli rejects the contemporary opinion that, events 'being controlled by Fortune and by God', men can have no influence upon them. For the modern reader this is an unpromising opening: Machiavelli is confusing the categories of explanation. But, as so often in his writings, the example he cites shows a solid realism beneath the deceptive flowers of rhetoric: an apparent intervention of *Fortuna* may be explicated as a change of circumstances no longer favourable to the effective working of the victim's 'innate character'. It is a long call from this insight into historical explanation, and its application in political action, to the case of the natural sciences, where neither 'free choice' nor 'character' has any relevance: a certain parallelism to Leonardo's notions of *forza* and *spirito* may not have escaped the reader's notice. Nevertheless it does constitute a clarification of the apparent paradox of an ordered universe in which at the same time elements of what have been called 'ultimate irrationality' cannot be excluded. This paradox was to dog scientific thinkers for a long time to come: it is much more *reasonable* to believe that heavy bodies fall more rapidly than light ones; but provided the resistance of the air is minimised they don't. Until this awkward fact had been accepted there could be no *science* of dynamics. The influence of Machiavelli's teaching in bringing about a climate of opinion in which men could come to terms with such paradoxes cannot be ruled out.

The fascination of Machiavelli is not restricted to the *morality* (or immorality) of his 'modern' *Realpolitik*. A typical child of the Renaissance, educated in the humanist tradition, he could not forbear to take 'one longing ling'ring look behind'. At least in *The Prince* his ideal is the military and political wisdom of the Mother that contemporary Italian rulers had forgotten or betrayed, Imperial Rome. But by the time *The Prince* was written (1513–14) the 'humanist' faith in ancient models of language and wisdom was fading, the science of warfare was being rapidly transformed, and even the fine arts were about to experience, in Kenneth Clark's phrase, a 'failure of nerve'. Indeed so far as Italy was concerned the Renaissance itself was coming to an end, and the form it took in the north-western marine countries was so different as to cast doubt on the appropriateness of using the same term to describe the change. Above all, the compass of the world and the kingdoms thereof had been enlarged beyond anything the ancients dreamed of. The southward extension of Africa had been doubled, the Indian Ocean could no longer be regarded as landlocked, the immense continent of America had been newly discovered. In all these new lands tropical or antipodean, men (without tails, and with heads above their shoulders instead of below!) were living in a state of vigour and natural luxuriance surpassing that of the bare north. Abundant evidence of the magnificence of eastern courts had corroborated a merchant's tale against the speculations of philosophers. Within a few years of Machiavelli's death (1527) the Roman genius for administration and land communications was shown to be rivalled by the Inca civilisation of Peru and by that paragon of orderly government, China, the latter stemming from the practical wisdom of a philosopher K'ung Fu-tsze, born more than a century before Aristotle. The tradition initiated in Greece, taken over by Rome, and beloved by the humanists, by which all mankind, other than those of this

supposedly unique tradition of civilisation, were lumped together as
'barbarians' was wearing a bit thin: the moderns could claim even
a great superiority over their revered forbears by virtue of the
knowledge of the vastness and variety of the world. Machiavelli's
delusion of grandeur, to be revived by cultivation of the ancient
arts of government, was out of date. His less well-known but possibly
more brilliant contemporary, Francesco Guicciardini, even if he
saw little future for the disorderly and fragmented nation that was
Italy, at least recognised the futility in profoundly changed circumstances, of seeking to model it on the Roman past.

However much—or little—Machiavelli may have been the founder
of *Realpolitik*, he failed signally to realise that henceforth the future
of a nation must be seen against a background of *Weltpolitik*. Even
before *The Prince* was written the demarcation of 'spheres of
influence' in respect of past and future oceanic discoveries posed a
problem on a scale such as Western Christendom had never before
had to face. The possibility of such a problem had in fact been
foreshadowed by the Portuguese colonisation of the Azores and
Cape Verde Islands; recognition of the problem was signalled by
the Treaty of Alcacovas (1479) and the papal bull *Aeterni regis*
(1481) which confirmed Portuguese sovereignty over all islands that
might be discovered south of the Canaries and west of Africa. This
E–W line of demarcation was a just and reasonable settlement in
the existing circumstances. By 1493 times had changed: Columbus
had unquestionably planted the Spanish flag on two large islands
far to the west of any that were likely to have been envisaged in
the bull and treaty; also he had reason for supposing the existence
of mainland to the *south* which, if it turned out to be the far coast
of Asia, as he believed, would have nullified the value of Vasco da
Gama's recently successful conclusion of the long and arduous
Portuguese struggle to achieve a sea-route to India. Recognising
how much was at stake the Spanish sovereigns induced the pope,
Alexander VI, to issue a series of bulls including the two famous
inter caetera, the earlier confirming the full sovereignty of Spain
over the recently discovered islands of Cuba and Hispaniola, the
later establishing a *meridional* demarcation line at 100 leagues (38°)
west of the *Azores*.

Such are the essential geographical facts. Their significance for
the future of international politics lay in a set of political facts,
the most important of which was that Alexander VI was a Spaniard,
Rodrigo Borgia, who, in the pursuit of establishing the secular power
of his son, Cesare, had become beholden to the Spanish sovereigns.
Later in the same year came a further bull, *Dudum siquidem*, virtually
giving to Spain *carte blanche* wherever and whenever her future
discoveries might take place. Realising that this outrageous 'gift',

extracted by political means, had shown that Spanish blood was thicker than spiritual infallibility, the Portuguese king John II bypassed the Holy See and sought to undermine its ruling by diplomatic confrontation with the Spanish sovereigns. The outcome was the Treaty of Tordesillas (1494) by which the papal demarcation line was shifted to 270 leagues west of the Cape Verde Islands in Portugal's favour. The technological superiority of the Portuguese navy had spoken louder than the throne of St Peter, louder even than the signatories could have been aware of: since by ruling that all future discoveries, *by whomsoever made*, east of the new line should belong to Portugal the treaty ensured that, while the islands discovered by Columbus would remain in the hands of Spain, the immense continental land area of Brazil would subsequently fall to Portugal.

This demonstration that the apportionment of territorial sway could be effected exclusively by secular powers revealed that the ideal of a united Christendom was no longer a political reality. Despite some inevitable lapses in an imperfect world this ideal had set the pattern of sovereignty throughout the Middle Ages; its foundation was the 'feudal' assumption that princes held their 'fiefs' from God, and since the pope was (and of course in the Roman faith still is) God's vicegerent on earth, his ruling as to the sovereignty of newly discovered lands was paramount. Once this ideal had been effectively replaced by the unspoken convention that princes had not only the right but the duty to exercise absolute power to ensure the 'vital interests' of their subjects, the need for a new regulatory code became an urgent necessity. The intractable nature of the problem postponed its solution—if indeed it ever has been solved—for at least a century, during which the degrading of the ambassadorial function from being a search for the means of maintaining peace to that of gaining an advantage over a rival state was furthered by means far more 'Machiavellian' than Machiavelli himself would have willingly countenanced. The demoralisation of international relations that characterised the sixteenth century may not have been the direct consequence of the Treaty of Tordesillas: such was the anarchy of the petty Italian states and the greed of the rising nation-states that the outcome might have been very much the same had the New World and its attendant problems never existed.

In a book published in Madrid in 1599 the instruments for the creation of the Spanish empire in America were represented as the 'sword and the compass'. More important than the geometrical compass shown in the picture was the magnetic. Despite characteristic claims by Renaissance writers the magnetic compass was an early medieval invention and sufficed for the 'discovery' of America.

The subsequent exploitation of the new continent, however, posed the problem of rapid transport of huge quantities of men and materials in both directions. To this challenge the application to navigation of ancient astronomical science, together with modern improvements in ship design, stood up fairly well. To meet the political challenge of achieving the most fruitful relations with the immense and varied population of the New World what was available was mostly *pseudo-*science: 'mostly', since a more intelligent application of Aristotle's theory of classification, based on essential rather than superficially striking characteristics, might have avoided much of the chaotic and senseless destruction that ensued.

In his recently published book *The Old World and the New* Professor J. H. Elliott writes: '. . . the discovery of America was important less because it gave birth to totally new ideas than because it forced Europeans to come face to face with ideas and problems which were already to be found within their own cultural traditions'. In the early decades the dominant ideology in terms of which the new continent was envisaged can be expressed as the expropriation of silver, slaves, and souls—each category involving a fatal lack of imagination and in the long run bringing some disastrous consequences to the expropriators. Silver, that 'oiled the wheels' of both commerce and artillery, ultimately contributed to a cost-inflation on an unprecedented scale and diverted attention from the enormous *real* wealth of the new lands. Slavery degrades both parties to the transaction and almost certainly stifles inventiveness. Of the good intentions of the majority of the missionary-priests there is no reason to doubt; but in the absence of any science of cultural anthropology other than the Aristotelian bifurcation of 'rulers' and 'meet to be ruled' the parochialism of the medieval church was invoked to justify the destruction of any supposed obstacle to the assimilation of the 'barbarians' to its own cultural pattern. St Paul (1 Cor. xiv.10–11, quoted by Elliott) knew better; but he was not a 'European'.

As early as 1552 there was published in Spain a damning indictment of his own countrymen by Bartholomé de Las Casas who, as priest and bishop, had spent a great part of his life studying the 'Indian' at first hand. About the same time as this published indictment Las Casas composed a remarkable work providing a theoretical justification for a more enlightened approach to the new problem of colonialism. From Aristotle's claim that 'Man is a *rational* animal' the Stoics, as Las Casas had learnt from Cicero, had inferred the possibility of a *ius gentium*, that is, a *universally* applicable law. From the evidence of high rationality, especially among the Aztecs and Incas, he inferred that there existed, in America, civilisations strikingly different from, but not necessarily inferior to, those of the

Old World. Unfortunately his work never achieved publication until the nineteenth century. In the ensuing half-century, however, enough information was disseminated from other sources to compel Europeans to question many cherished beliefs: Jean Bodin began to doubt the existence of a Golden Age whose alleged wisdom was still being sought by some of his contemporaries; but the notion that continuous change necessarily implied progress the sceptical Michel de Montaigne seriously doubted.

It was also Jean Bodin who first put forward a reasoned case for the connection between the continued cost-inflation of the sixteenth century and the increased amount of coin brought into circulation as a consequence of the importation of American silver. This is an outstanding example of the too ready acceptance of an observed change as 'the' cause of a dramatic divergence from the 'normal' course of life. More refined statistical studies in recent times have cast doubt on the possibility, at least before mid-century, of so large an effect in view of the small amount of imported silver when compared with the probable output of the central European mines (p. 40). Similar studies have shifted the 'blame' on to other factors of change, notably the growth of population outstripping improvement in agriculture. It is in any case emphatically a matter for experts. For our purpose the correctness or otherwise of Bodin's hypothesis is of less importance than the fact that he formulated it.

In the last years of the century came the comprehensive survey of the 'Indies' by José de Acosta which, together with Harriot's much smaller monograph on 'Virginia' (p. 109), provided a model for some years to come. They were followed after 1600 by numerous others.

If we ask to what extent a 'science' of politics was possible at the end of our period, the answer must be, none. It was an age not only of rapid change in the systematisation of collateral enquiries but also one of intellectual confusion. If it be conceded that political theory must take account of the contemporary state of historical understanding and of the world-view consistent with the state of natural knowledge, it must be remembered that both the scientific and historical (p. 172) 'revolutions' were by then under way. The traditional foundations in each sphere were crumbling, but there were as yet no others from which new forms of order and sovereignty could be envisaged. For political theory, as for historiography, the introduction by Joseph Scaliger (*De Emendatione Temporum* 1583) of critical methods of establishing a chronological reference frame based on historical rather than on mythical evidence was highly significant. The time-scale was of course far too small and was to remain so until geological evidence made nonsense of the whole biblical concept of creation.

INNOVATION AND METHOD IN MEDICINE

On 7 August 1495 in an edict issued at the command of the Emperor Maximilian I there appeared the first printed reference to a disease, 'called the evil pox never previously existing nor heard of in the memory of man'. Within three years there had appeared at least ten tracts directly concerned with this disease that was already referred to most commonly, though not exclusively, as the *morbus gallicus*, or the equivalent in each vernacular. The authors of the tracts were physicians attached to the persons of princes or to the corporate bodies of cities, independent practitioners, or university teachers. In one particularly significant case—that of Nicolò Leoniceno—the tract was based on a disputation at the University of Ferrara and is worth somewhat detailed study, as well for the character of the author as for the interest of its subject-matter.

Nicolò was born in 1428 at Lonigo near Vicenza, hence the epithet 'Leoniceno'. After study and teaching at the long-established universities of Bologna and Padua he settled in 1464 at Ferrara where the almost equally famous Ugo Benzi (d. 1439) had in his last years established the university's reputation as a centre of medical teaching rivalling that of the older universities. Hugo's reputation was admittedly based rather on his eloquence and skill in worsting a rival in disputation than on any solid contribution to Medicine as such, but his clarification of its *method* marked an important stage in the subsequent emergence of Medicine as an art based on rational procedures rather than on the authority of allegedly successful practitioners (p. 86). Helped by the mastery of Greek, which the influence if not the direct teaching of Guarino (p. 22) made possible, Leoniceno was able to bring a critical eye to bear on the texts of the sources to which disputants were wont to appeal. One of the most

important of these was the elder Pliny's *Natural History*, the main
source of medieval 'knowledge' of the world of nature, which,
though of course written in Latin, drew heavily on the Greek
sources available to Pliny. Leoniceno's critical examination of the
text revealed a large number of linguistic and terminological con-
fusions; but towards the alleged facts to which these referred he was
less critical; nevertheless the establishment of a reliable text enabled
his contemporary, Pandolfo Collenuccio, to take Leoniceno himself
to task for not having paid more attention to the 'facts'.

Leoniceno's *Booklet on the epidemic commonly known as the
French Disease* displays both the strength and the weakness of the
humanistic-academic approach to Medicine. With his intimate
knowledge of the Greek and Latin classics he was able to form a
reasoned assessment of the likelihood of its being an entirely new
disease, or at least previously unrecorded: Pliny, erroneously as he
believed, had made a similar suggestion about a kind of ringworm
(*lichen*). Leoniceno's comment is worth quoting:

When I reflect that men are endowed with the same nature, born under
the same skies, brought up under the same stars, I am compelled to think
that they have been always afflicted by the same diseases; neither can I
comprehend how this disease has suddenly destroyed our age as none
before. For if anyone thinks otherwise than I do what, I ask, would he
say that this is other than a revenge of the gods? For if the laws of
nature are examined they have existed unchanged on countless occasions
since the beginning of the world. Wherefore I am prepared to show that
a similar disease has arisen from similar causes also in past ages.

Thus far the analysis is in keeping with the 'scientific' approach
of the Hippocratic school of Medicine—critical reference to similar
visitations in the past, cautious in falling back on the easy alternative
of complete novelty, affirmation of one of the essential principles
of scientific investigation—similar effects are the consequences of
similar causes. But what of the 'revenge of the gods'? To have
excluded this aspect of disease at that time, or for centuries thereafter,
would have been almost unthinkable; the plural form was probably
a humanistic mannerism similar to the confusion of pagan and
Christian symbols in religious art. And when we read on we find
a clear distinction (as in the Hippocratic work on the 'Sacred
Disease'—epilepsy) between the 'theological' truth, outwith the
purview of the 'scientist', and the proximate (*proximiores*) causes
with which alone Leoniceno would concern himself. These are the
powers of nature—'in the year in which the disease began to spread
(*pululare*) the waters arose all over Italy'—to this abnormal state
he ascribed the onset of the disease: to us implausible but not
'unscientific'.

The weakness of this approach is most apparent in the reverence shown towards the classical writers, especially in respect to the 'humoral' theory in which every disease was held to be an expression of the disturbance of the 'harmony' of the 'humours' constituting the healthy body. These—phlegm, blood, yellow and black bile— were characterised by the different proportions of the Aristotelian elements—earth, water, air, and fire—which in turn were formed by the four possible combinations of the qualities wet, dry, hot, and cold. The proportions of the 'elements so mixed in him' determined a man's temperament—phlegmatic, sanguine, choleric, or melancholic. Consonant with this theory Leoniceno claimed that the appearance of pustules on the *pudenda* is explained ('as Galen said') as a consequence of the excessive warmth and moisture of those parts. Blinded as he was by his elegant, plausible, comprehensive, and almost certainly totally erroneous, theory of the nature of all diseases Leoniceno did not even entertain the possibility of transfer of 'contagion' in the sexual act; that this was not quite so purblind as has sometimes been made out must be conceded in view of the fact that cases of non-genital infection were by no means unknown.

If ever a 'breakthrough' was needed in science it was surely here: it was effected by an 'angry young man' (Paracelsus, p. 87) who claimed to have sat at the feet of Leoniceno. But more widely influential was the less brutal invective of another undoubted pupil, colleague, critic, and successor of Leoniceno, Giovanni Manardi (p. 74), who put the humanistic teaching in the most favourable light and at the same time pointed the way towards its critical revaluation; to him Paracelsus may have owed more than to Leoniceno.

Where Leoniceno had been cautiously critical, and largely in respect of the comparison of texts, Manardi was far less compromising: for him it was not enough to collate different meanings of, for instance, 'rhubarb' among the ancient authorities; these must be tested not so much against one another as against specimens of the plant itself. The same holds in respect of diseases: it is indeed essential to establish a standard meaning of a term, but to rest at this level is to be a 'textbook doctor—one of those who having found its [*sc.* the disease's] name looks up the remedies appropriate to it in their books'. 'True' doctors on the other hand will not be content until they have discovered by the appropriate method (pp. 91f.) the cause of the disease such as will suggest effective treatment. 'They will make books their servants and not follow them blindly as masters.' Nor does his scepticism end with words; greatly daring, as Manardi himself admitted in a letter dated 1518, he refused to accept the decision of any man as final. Three years later in the dedication of a book to his medical students he makes the point in the striking exhortation: 'In return for these my labours I shall nevertheless require of you one undertaking only, namely that in respect of my comments [*sc.* on the pharmacopeia of the Arabian, Mesue] you will defer to the authority of no man, be he dead or among the living, more than to reasoning and truth.' This was 140 years before the Royal Society of London adopted for its motto *Nullius in verba* (p. 151).

Manardi's views on the 'French pox' were set forth in a letter written at Mirandola, where he was acting as personal physician and collaborator with the younger Pico in the posthumous edition of the works of the latter's uncle (pp. 149f.). Though his constructive thought is constrained, as was that of all his academic contemporaries by the all-embracing humoral theory, he rejected out of hand the opinion put forward by a physician, Simon Pistor, that the pox was literally 'disastrous'—the effect of an occult emanation from the heavenly bodies. Though, as a humanist, Manardi was of course highly critical of all the Arab writings, he nevertheless went on to say that if Pistor had read his Avicenna carefully he would have seen that the eminent author of *The Canon of Medicine* had represented the heavenly bodies as exerting a *universal* force and not one of an occult or specifically 'evil' nature: there was no feature of the *morbus gallicus* that couldn't be accounted for by an abnormal (though not *unnatural*) change in the balance of the humours.

In the same letter Manardi calls in question the excessive time spent by medical students in gaining a mastery of astronomy such as would enable them to choose propitious days for purging, blood-letting, and the administration of drugs associated with certain planets or signs of the zodiac. Towards the end of his life in Ferrara, strengthened in his doubt of the infallibility of the Greeks by the refutation of their prejudices against the habitability of the Torrid Zone (p. 74), he felt, 'compelled to part company with Galen, and not from him only but from those whoever they may be who strive to smear and corrupt with astrological superstitions the most distinguished and chaste art of Medicine'. Bold sentiments in an age as obsessed with astrology as is ours with sex! But Manardi's was not just a generalised negative prejudice; it was backed by sound clinical insight: 'the pulsation of the veins is a surer guide than the configuration of the heavens'. To a large extent his words fell on deaf ears: for nearly half a century thereafter the most successful textbook (p. 94), in many respects critical and well ordered, claimed that a knowledge of astronomy was an essential part of the physician's training.

Consonantly with the natural history of Aristotle the 'cure' for the *morbus gallicus* should be found growing in the country where the disease itself—if it was in fact one previously unknown in Europe —was endemic, namely, the 'Indies'. Among the indigenous trees, sought not only as potential timber but also as a source of natural dye, was *Guaiacum officinale* (*Lignum vitae*), a concoction of which has a bitter taste and would probably exhibit some degree of fluorescence. Possibly because of these properties and the perverse belief that the more nauseating the treatment the greater the therapeutic potency, this preparation 'caught on' as an alternative to the far more dangerous (but also far more effective) treatment with mercury. A welcome boost came from the celebrated knightly humanist, Ulrich von Hutten, whose booklet in praise of the miraculous drug ante-dated his death from the disease in 1523 by four years. Welcome, since the merchant-banker house of Fugger in Augsburg exploited their monopoly for the import of the guaiac tree by selling the bark at an exorbitant price. In this enterprise they were aided by the clinical observations made in Spain by Nicholas Pol, physician to the Emperor Maximilian. It would be unjust to infer from his favourable report that he had acted in an unprofessional manner, since no one was aware at that time that the primary signs and symptoms of syphilis disappear spontaneously, and that the appearance of the secondary may be long delayed.

The mercurial treatment on the other hand received in 1529 an even more spectacular 'advertisement' from a quarter hardly likely to commend itself to the more 'classical' physicians; this was the book called *The French Disease*, one of the few works printed in the lifetime of its author, 'Philippus Theophrastus Bombast von Hohenheim Paracelsus genannt', as he was commonly described by his editors. He was born at Einsiedeln in Switzerland, son of a physician to the local hospice and his wife who held a rather superior position therein. His father came of a 'noble' family that had fallen on evil days—the name 'Bombast' being a corruption of the earlier 'Baumast' having no connection with the English word 'bombast'. On the early death of his mother, who seems to have had a great influence over her son, he accompanied his father to a new appointment as physician to the Fugger silver mines in Carinthia. There the boy learnt from his father the association of all natural phenomena—minerals, vapours, springs, plants, etc.—with the practice of Medicine and, perhaps more important, the changes that could be wrought in them by man's ingenuity. One of the few original discoveries made later by Paracelsus was the association of certain local sources of drinking water (deficient as we now know in iodides) with the locally prevalent disease of goitre. His subsequent nomadic life is fairly well documented; his claim to have graduated in 'both medicines' at Ferrara (for his attendance at which there is some circumstantial evidence) not at all; his first academical appointment in 1527 as municipal physician with the right to lecture at the university in Basel is authenticated in almost every significant particular.

From the 'chair' at Basel, by then one of the leading medical schools in Europe, Paracelsus poured forth a stream of invective against the 'academic' physicians and their tutelary deities, Galen and Avicenna, some of whose books he hurled on to the bonfire of the usual St Johannisnacht celebrations. Unrestrained as his language certainly was—a fair sample of it was posthumously published under the title *Paragranum*—and, to add insult to injury, delivered in German spiced with Swabian dialect instead of in the customary Latin, his attack differed from those of some of the 'protests' of our time in having a sound basis of criticism and in offering a constructive substitute. His main targets were the hopelessly vague humoral theory of disease and the composite concoctions of plants—often containing a dozen or more ingredients—that were the chief armoury of the Galenists in their efforts to effect a cure. To replace the humoral theory he developed (in a book called *Volumen Paramirum*) an admittedly fantastical scheme that nevertheless had the virtue of ascribing to each disease the introduction into the body of a specific *external* cause or 'poison'. The alleged celestial origin of many of these cannot be dealt with here, since the notion was the product of the persistent concern with magic that, interacting with the more 'scientific' ideas and procedures of the Renaissance, can be put in perspective only at some length and in a general historical setting (Chapter 11). Suffice it to say at this stage that, though to a large extent misconceived as to details, the pathological insight of Paracelsus effected in some quarters a 'breakthrough' without which, delayed as it was in its effects, it is difficult to see how any fundamental advance in Medicine could have come about.

The change that Paracelsus advocated in respect of therapeutic measures, though more generally counted as his chief claim to a prominent place in the history of science, was not quite so original. Instead of the Galenic polypharmacy to which he gave the untranslatable name of *Suppenwust* ('soupy muck' is probably a fair rendering) he introduced the much wider use of mineral substances, many of which, such as preparations of iron and antimony, he was the first to prepare. The novelty here was in urging their *internal* use as well as the traditional external application, and in drawing the highly suggestive analogy between the working of the physician (as well as the human body itself), the metallurgist, the baker, the farmer, with that of the alchemist. In the late medieval centuries a few men had indeed claimed that *quintessences* prepared by alcoholic extraction of all kinds of materials had a curative power far greater than that of aqueous concoctions; but again it was an *insight* into a universal character that marks out Paracelsus, despite his destructive and often incomprehensible rhetoric, as a genius who let loose into Renaissance Medicine a power for good and evil that had repercussions down to and including our own time.

Many of the 'cures' effected by Paracelsus were probably due to 'suggestion' (despite his unbridled tongue he was a deeply religious man); many must also have been the consequence of luck rather than understanding; but there is now little doubt that he used ether as an internal medicine. The effect would have been that of an analgesic similar to that of 'chlorodyne' (mainly chloroform) used much later in the treatment of painful internal maladies such as cholera. The relaxation thus achieved may have hastened the passing of the disease; an overdose may equally have hastened or even provoked the passing of the patient: but the treatment can hardly have failed to enhance the reputation of the physician. There is considerable evidence that Paracelsus was led to the discovery of this *arcanum*, as he would have called it, by observing the soporific action on domestic fowls of the product obtained by distilling 'oil of vitriol' with 'aqua vitae', to which incidentally he was the first to give the name 'alcohol'. A clear account of the preparation of ether was, however, first given by the botanist, Valerius Cordus, in a book that appeared posthumously under the editorship of Conrad Gesner.

While Paracelsus was 'spoiling for the fight', launched at Basel in 1527, the Doctors of Medicine emerging from the 'humanistic' medical schools had been making enlightened contributions to those branches of Medicine against whose conservatism Paracelsus had unleashed his wrath. Another pupil of Leoniceno and Manardi, Antonio Brasavola, had somewhat reluctantly betaken himself 'to the high and inhospitable alps' to study plants in their native habitats rather than in books. With the knowledge thus gained and with some resort to actual trials he issued a series of works in which the time-honoured 'simples' and other more complex preparations were critically passed in review and in many cases rejected. His work would have been carried much further by the German Valerius Cordus had he not died at an early age in Italy as a probable consequence of exposure in the course of prolonged field-work. He did however leave to the city of Nürnberg sufficient material to form the basis of the first official pharmacopoeia.

Of more direct consequence to the progress of Medicine was the appearance in 1514 of a small book on *Some hidden and astonishing causes of disease and cures*. Antonio Benivieni, the author, is an example of a man capable of combining purely literary studies in the classical tongues (in this field his brother, Girolamo, was even more famous) with the daunting manual operations of the *post mortem* slab. Despite his fame as a physician the manuscript record of his three hundred autopsies was printed only five years after his death (1502) and then only in a very much contracted form. This version was reprinted in 1528 (Paris) and 1529 (Basel) but thereafter only after a lapse of over fifty years. In 1581 the physician-botanist, Rembert Dodoens (p. 110), included it in a collection of 'rare medical observations' culled from several authors. In his editorial introduction Dodoens emphasised the 'change in manners and customs from former times whereby human bodies might be opened'. For more than half a century after his death the outstanding significance of Benivieni's method and foresight was probably lost on the majority of physicians; it is perhaps even more remarkable that when Rembert Dodoens published his appreciation nearly forty years had passed since the appearance of Vesalius's epoch-making work (p. 106). Misleading and unhistorical as such ascriptions usually are, it is difficult to withhold from Benivieni the title of 'Father of Morbid Anatomy'.

A word on the 'change in manners and customs' may be necessary to put the practice of human anatomy in proper perspective, if only to do away with the still prevalent myth that no advance in anatomy had been possible owing to the intransigence of 'the Church'. Whereas Galen—superb anatomist though he was—had been hampered by the contemporary disapproval of dissection of the human body, upon which the Koran had later put an absolute ban, the Christian Church had, at least from the thirteenth century, merely taken upon itself the duty of *regulating* the practice. The University of Bologna—a city at times under direct papal hegemony—was a pioneer in the holding of regular, if infrequent, public anatomies. The first textbook of anatomy to provide any advance on Galen, to the extent that it was based on the dissection of the *human* body, instead of on the Barbary Ape, was compiled at Bologna by Mondino de' Luzzi in 1316 and was not displaced until well into the sixteenth century. The comparatively minor advances in the subject during the Middle Ages must be ascribed rather to lack of enterprise in the medical schools than to proscription by the Church. The most charitable explanation of the popular myth to the contrary is that academical dissection was confused with the ecclesiastical ban on the boiling of bodies of dead crusaders to enable their bones to be returned to their homelands—surely a not unreasonable veto.

There is reason to believe that Benivieni was not the first to carry out autopsies to discover the cause of death where foul play was suspected; but this was coroner's work rather than medical research. For contrast let one of Benivieni's cases speak for itself: 'A young man otherwise well-seeming suffered from severe griping of the bowels. Though there was no recognisable cause he ultimately died, wherefore the body having been incised so that we might learn the causes of so great an ill, and all the viscera thus uncovered being as it seemed to me in a healthy condition, I ordered the intestine to be removed. When my assistant cut this out with a scalpel he came upon a large abscess from which, when it had been cut, there flowed a large quantity of black and fetid matter similar to ink (*encausticum*). Whence we recognised the evident cause of death; for the black bile flowing thither had gradually produced an abscess of this kind.' The acceptance of the (imaginary) 'black bile' as the ultimate cause reveals no advance in respect of physiology; but the demonstration of an *internal* morbid *structure* was a promising beginning. Another case, that of 'Antonius Brenus', is illuminating in other respects. First, he was a relative of Benivieni's; second, he was suffering from a wasting disease (neither food nor medicine achieved 'coction' in the current Aristotelian terms): third, Benivieni claims to have carried out an autopsy of the dead body *publicae utilitatis gratia* ('in the public interest'). It revealed a purely structural deformity as a sufficient cause, namely an obstruction of the gut. It should be added that not all Benivieni's subjects got as far as the mortuary slab; many seem to have recovered after relatively minor investigation had revealed the 'hidden causes'.

How far Benivieni's pioneer work promoted the striking rise of the anatomy schools in the sixteenth century it is hard to say; but it is clear from Dodoens' tribute (p. 90) that well before the end of the century it was realised that progress in medical understanding could not be brought about until an exact knowledge of the 'working parts' of the *normal* healthy human body had been achieved. This separation of anatomy as a science in its own right independent of any application to the art of Medicine was to a large extent, though not quite so exclusively as he and subsequent historians have made out, the work of one man, Vesalius, of Brussels. Since the prologue to this great drama took place in the Low Countries and Paris it will be appropriate to postpone its further consideration until the culmination of the Italian humanist tradition in medical science and philosophy has been dealt with.

As the sixteenth century wore on, the focus of Italian medical science moved from Ferrara to Padua in the mainland territory under Venetian rule. Padua had for a long time been what Shakespeare at the end of the century was to call a 'Nursery of Arts'; and among these arts the one for which Padua had been most famous was a critical study of the logic of Aristotle, though carried out, it must be admitted, mainly in the 'higher' Faculty of Medicine. This study was initiated by Pietro d'Abano (Abano's mineral springs bordering the Euganean Hills near Padua have been famous since Roman times;

it is now a fashionable 'spa') whose knowledge of Greek, unusual at the beginning of the thirteenth century, enabled him to expound a theory of Medicine in which the advanced logical work of Aristotle (the *Posterior Analytics*) played a basic part. In his principal work (generally referred to as *Conciliator*, written in 1310, first printed in 1496) he applied his great and wide knowledge to the laudable task of trying to resolve the dissensions between the philosophers and physicians. The only section of this large book we are concerned with is Pietro's enquiry into the question as to whether Medicine is a 'science' or an 'art'—a question that still arouses interest and even tempers. Exposition of the course of this enquiry necessarily involves a few logical technicalities; but without some appreciation of these it is impossible to understand what the medical school at Padua was about. The problem was indeed posited a long time before the period we are concerned with, but consideration of it persisted at least till the end of the sixteenth century; and it was in such an atmos-phere that Galileo and William Harvey were trained to think about the problems of the 'new sciences'. There is *explicit* reference to this in their works. And we might add that some recent high-sounding claims to a 'breakthrough' have been made on evidence whose logic would have been rejected out of hand by the Paduan doctors.

Shorn of detail the problem of scientific knowledge may be put as follows: 'We think that we know a thing unqualifiedly and not in a sophistical and accidental manner when we think we know the cause of that fact and that it could not be otherwise.' Thus far Aristotle, expounded by Pietro d'Abano. The difficulty arises from the fact that in Medicine, as Benivieni emphasised, the 'cause' is usually 'hidden'. In order that our search for the hidden cause may have a rationally based conclusion we must have recourse to another form or aspect of knowledge, that is, of what is more knowable and certain to *us,* namely, the 'facts' of experience. In other terms, from empirical knowledge of the observable circumstances we infer, by the method known as 'resolution' (but of course not *rigorously*) a general proposition from which the facts can be *rigorously* deduced; this latter was called the 'method of composition'. All this was unexceptionable; it was the method employed in answer to his critics *explicitly* by Harvey, who had studied in Padua, *implicitly* by Iohannes Kepler, who hadn't, in establishing two of the major components of the 'Scientific Revolution'. What, in contrast to Kepler and Harvey, these optimistic medievals didn't realise is that the establishment of the *totality* of *relevant* facts may be even more exacting than the intuiting of the universal proposition from which they necessarily follow. We shall return to this question in the last chapter.

Unfortunately this promising start became enmeshed at an early

date with the exposition of a famous theoretical work by Galen
called the *Art of Medicine* (also the *Ars parva*). The art of *teaching*
thus became confused with the art of *investigation*. So that when the
great masters of Ferrara—Leoniceno and Manardi—came out as a
foil to those of Padua an enormous amount of time and printers'
ink was expended on refining an issue that to a modern reader appears
to have been largely a 'semantic confusion'. John Caius, who wrote
a tract on the subject, with characteristic Cambridge scepticism,
came to the same conclusion; but he was ready to admit, as were
many of the foremost physicians of the later sixteenth century, that
he was greatly beholden to the lectures of Giovanni da Monte—
professor of Medicine at Padua at the time that Caius was there
expounding Aristotle in Greek. If people wanted evidence that da
Monte's method led anywhere, Caius suggested that they should
refer to his own tract on *The English Sweats* that had been modelled
on the method.

The widespread influence of da Monte's teaching is evidenced
by the fact that two separate collections of his works were made
after his death by physicians of the medical school at Breslau in
Eastern Germany. One of these is of exceptional interest since it
retains the actual words of his opening lectures on method, so far
as these could be discovered from manuscripts and the testimony
of some who had attended them. To the two methods (resolutive
and compositive) da Monte added the *divisive* by which a complex
problem is broken down into manageable components. This bears
a strong resemblance to the insistence by Descartes on the establishment of *clear* and *distinct* ideas as essential preliminary to any
true knowledge. In his correspondence (though not in his *Discourse*,
that despite its title has hardly anything to do with *method*) Descartes
uses the terms 'analysis' and 'synthesis' in senses hardly distinguishable from 'resolutive and compositive methods'. Being Descartes he
did not acknowledge his sources; but to be fair to him it is possible
that by then these methods had become so widely referred to as to
render acknowledgment superfluous.

The modern reader is struck forcibly by the degree of *logical*
sophistication that the young men in da Monte's audience are
assumed to possess; but the great majority would have been well
grounded in the Faculty of Arts. It was freely admitted that he did
not spare them; there were some who grumbled at his teaching
being too 'theoretical'. Maybe it was; but what cannot be denied
is that though he may not have been the first to illustrate his lectures
by means of 'case-histories', yet by 'exercising his students in the
resolutive and compositive methods within the hospital of San
Francesco' he opened up a new dimension of medical teaching;
and, since for a century thereafter the majority of 'natural scientists'

were graduates in Medicine, such teaching must have made an impact on scientific thought in general. If this has not been taken sufficient account of by historians it is a consequence of the artificial separation of the history of 'science' from that of Medicine. Moreover the seemingly endless disputes as to whether the teaching of science should aim at 'facts' or 'principles' might have been greatly curtailed had the disputants read da Monte's brief tract on the subject. Teaching, he claimed, must unavoidably be carried on mainly by the compositive method, that is, deductively from generalisations to their individual consequences. To learn by the sole use of the resolutive method would be too difficult for beginners. But the *exclusive* reliance on either method would involve the impossibility of any *knowledge* of the natural world.

In the early decades of the sixteenth century the term *methodus* was very much in the air—Thomas Linacre (p. 104) had in 1519 translated from the Greek a famous treatise by Galen under the title *Methodus medendi* (Method of healing) and Erasmus in the following year published a work in theology whose title began with *Ratio seu Methodus. . . .* Nowhere to be found in classical Latin the word, at least in the sense intended by Aristotle and Galen, was a creation of the Renaissance. The Greek word μεθοδος means literally 'after the way', a sense brought out in the expression *via et ratio* employed by Cicero wherever the Greek μεθοδος was appropriate. In 1541 Leonhart Fuchs (p. 63) changed the title of his little *Introduction to the art of Healing* to *Method or theory . . .*, the work being enlarged and revised at the same time. In the subsequent editions, further enlarged to include a good deal of surgery, the title changes but the 'methodising' of Medicine is apparent in the distinction between 'theory', now for the first time called 'physiology' in something like its modern sense, and 'practice'. In his titles Fuchs preferred *Institutiones* to *Physiologia*, thus giving rise to a custom exemplified in the great medical school of Edinburgh where the two chairs of Medicine, as distinct from those of anatomy, gynaecology, etc., were known respectively, at least until the beginning of the nineteenth century, as 'Institutes' and 'Practice' respectively.

This flood of 'textbooks' from his pen, as compared with the sparse editions of the *Consultations* of Manardi and the two collections of da Monte issued only posthumously, shows that though well versed in Latin and Greek Fuchs was moving out of the 'humanist' tradition of Medicine to provide a more systematic course than could be obtained by the student from the scattered works of Galen and the stupendous tomes of the Arabs, Avicenna and ar-Razi. Before following Fuchs into this more 'modern' style of academic Medicine we must, however, take note of one of the most striking figures of the declining Italian Renaissance.

As he enters the Piazza dei Signori in Verona the visitor's eye will be held by the lovely Palazzo del Consiglio built probably by Fra Gioconda about the time when Girolamo Fracastoro, whose statue stands on the arch to the left of the building, was born. His wide interests and attitudes, combining a degree of dilettantism and Roman detachment from the academic market place with a very modern-looking theory of epidemic disease, justify more than a passing notice of his work, which appeals to a very wide audience. To students of Italian literature he is known as the author of lengthy poems in Latin hexameters on poetry, intelligence, etc.; to historians of astronomy as the inventor of an ingenious but untenable revival of the Greek homocentric theory of planetary motions; to medical historians as the originator of the name 'syphilis', the name in fact of a youth described in a lengthy poem as having been struck down by Apollo with the by then only too well-known *morbus gallicus*. The last-named work provides a classical account of the signs and symptoms of the disease; but the name did not 'catch on' until the nineteenth century, *lues venerea* having replaced *morbus gallicus* during the later part of the sixteenth century. None of these works, however, equals in originality those written towards the end of his life on *Sympathy and Antipathy* and *Contagion and Contagious Diseases*.

Though occasionally taking part in disputations Fracastoro held only a very brief university appointment, preferring the rather isolated life of his villa some miles from the city. This detachment may have accentuated the uneven quality of his thought. He displays both the weakness and the strength of the humanist tradition: the former in his willingness to model so much of it both in style and content on the classical authors, notably Virgil and Lucretius; the latter in his critical grasp of those sources and in his selection of those that gave promise of more soundly based knowledge of 'the causes of things'. Relative isolation from the academic climate of opinion may account for the fact that his theories, consonant with experience (though not *experiment*), redressed the balance against the suggestive but unverifiable magico-mystical (neo)platonising of Marsilio Ficino then exercising widespread influence (p. 152). Neither in *Sympathy and Antipathy* nor in *Contagion* is there any striking originality: the extension of the manifestation of human 'emotions' to inanimate objects had a long tradition. Also the *seminaria* (in *Contagion*), to whose dissemination he attributed the appearance of plague and *morbus gallicus*, should not be identified with 'microbes'; they bear a closer resemblance to the 'atoms' of Lucretius, but are endowed with a kind of 'catalytic' power to stir up the 'humours' in living, or induce putrefaction in recently dead, bodies. Fracastoro's 'transitional' attitudes are likewise to be discerned in his distinction between primary and proximate causes: while the former, being from God, may be spiritual in nature, the latter are material. Also, despite his tract against reliance on astronomical portents determining 'critical days' for medical action, he was ready to admit, as had been Manardi, that the stars though immaterial might act on living creatures through the medium of material objects.

It is difficult to assess the influence of Fracastoro's theorising on the practice of physicians. There are many indications, notably in the works on the plague by the Italian Alexander Massaria, of a more realistic response to the periodical visitations of this scourge that swept across Northern Italy during the sixteenth century. Fracastoro's conclusions provided strong, though unfortunately not decisive, support for the pragmatic 'quarantine' regulations imposed by numerous municipalities from a time much earlier than his writings. Massaria and the municipal 'medical officers of health' may have been converted from the more orthodox medical theory of 'miasma' similar in form to that put forward by Leoniceno (p. 84) on the origin of the *morbus gallicus*. In the case of Massaria there is more than a hint that he at least knew of Fracastoro's emphasis on spread by contagion, even if he had not fully understood the suggested mechanism.

On the therapeutic side the sixteenth century was characterised by two innovations—the great hopes, only moderately realised (quinine was of the next century), of new drugs from the 'Indies'; and second, the growing dispute as to the relative value of the traditional 'Galenicals' (of plant origin) and the 'wonder drugs' of mineral origin advocated by Paracelsus. Towards the end of the century the latter reached the frenzy and bigoted folly of 'ideological warfare'.

The last remaining innovation of which there is room to speak is in the field of surgery which, it should be emphasised, embraced all treatment of the sick and injured by external ointments, dressings, etc., as well as operative procedures, whether external or internal. In both these areas new problems had been posed by the huge increase in the employment of gunpowder. Though this had been introduced into warfare before the end of the fourteenth century, there is little that can be gleaned about its medical impact before the printed book began to disseminate methods of treatment just before the end of the following century. In any case the main uses of gunpowder before that time had been to create noise and shatter masonry. The introduction of mobile artillery and, later, manual firearms posed two new problems: the extraction of shot from flesh and bone, and the largely theoretical query as to whether the bullet became 'poisoned' or 'purified' by the fire of the charge. Since the cautery had largely replaced the knife in the progressive surgery of the Arabic practitioners, the 'purification' by the explosion was by many regarded as rendering the shot no more lethal than the surgeon's instrument. It was about a half a century before the English surgeon, Thomas Gale, gave positive evidence that the heating effect of the burning powder on the bullet is far below that of the cautery.

In the technique of general surgery no fundamental change appeared during the Renaissance. Contrary to the still too prevalent views about the stagnation of the healing art in the Middle Ages the practice of such men as Theodoric of Bologna, William of Saliceto, Henry de Mondeville and Guy de Chauliac showed marked advances on both Greek and Arab practice. It may be significant that nearly all the leading figures had studied in Northern Italy or Montpellier (p. 105) where the boundary between Medicine and Surgery was not so sharply drawn as elsewhere and at other times. It might be further argued that the greater emphasis on literary elegance and mere erudition in the *quattrocento* accounted for the fact that the surgeons of the thirteenth and fourteenth centuries would have had little to learn from their successors. But, important as this question undoubtedly was from the point of view of the social relations of science, it is far from simple and not to be decided by ideological preferences.

The almost ceaseless large-scale wars of the sixteenth century increased the importance of military surgery. In those countries where it had been the custom for the physicians to regard the surgeons as an inferior caste the balance was shifted somewhat to the advantage of the latter; among those who took a notable part in this shift were several who first achieved celebrity as military surgeons. Since the stage for this change was set principally in France and England, involving new relationships with government, the matter can be put in clearer perspective in the next chapter, devoted to the Renaissance in those countries.

D

8

SCIENCE IN THE EMERGENT
NATION-STATES

'To speak candidly about this our age, the disciplines and arts that had been buried nearly twelve hundred years ago and had really fallen into a state of complete extinction have now completely revived and regained, I might even say surpassed, their former splendour, so that this age need envy that former time in hardly any respect.' Men of very varied attainments, the writer went on to point out, had created 'many magnificent works that yield nothing to those more ancient ones whose fame is on every one's lips'. And not only was this true, he emphasised, in respect of the fine arts and letters but also of military engines, fire bombs, printing, and the discovery of new lands.

It is profitable to compare this manifesto with that of Alberti quoted at the beginning of this book (p. 17). It must evidently have been written later than Alberti's since no really 'new' lands were discovered until about fifty years after the latter's encomium. It was in fact written a century later, and almost a century after the introduction of printing, the only triumph of human ingenuity, other than military novelties, that the author chose to mention. Yet he was no hack writer but Jean Fernel, the most influential physician in France, if not in Europe, during the second half of the sixteenth century; and it was he who had also made the first Western measurement of an arc of meridian. Allowance must indeed be made for the fact that he was addressing the French king Henry II in the dedication of his book *On the Secret Causes of Things* and Henry was hardly likely to be a very enlightened critic. But it was a very learned (if dated) work in which Fernel would surely have taken some care to avoid appearing so sadly behind the times as the tenor of the dedication suggests that he was. The truth seems to be that

French cultural development *was* behind the times; fifty years previously the same sort of nonsense as that with which the passage opens might have passed muster in Italy but hardly in 1547—four years, we may remind ourselves, after the publication of those epoch-making books of Copernicus (p. 116) and Vesalius (p. 106). We must not, however, 'draw up an indictment against a whole nation' on the evidence of a single document.

If we cast our eyes back over the period that we have attempted to assess in respect of progress in natural knowledge we are struck at once by the very small number distinguished in this field in France as compared with Italy and Germany. Setting aside universal geniuses such as Leonardo da Vinci, of which more than one in a century is itself a comparatively rare occurrence, there is not a single name in France from the early *quattrocento* to the time of Fernel to be put in the same class as even Toscanelli, Nicholas of Cues, Regiomontanus, Dürer, Manardi, Paracelsus. None of these, it must be admitted, stand out prominently in a history of *science* as commonly understood; but in an age of the 'transvaluation of all values' they, and many more that could be named, saw further and with surer vision than their contemporaries into what was to become natural science. In France we should look in vain for anyone of which the same could be said.

What of the remaining European countries? In the absence of geographical barriers such as the Alpine massif that demarcates Italy almost wholly from Germany and France, in turn demarcated from the Iberian Peninsula by the Pyrenees, it is misleading to attempt to delineate too closely 'national' boundaries applicable to the sixteenth century. But with this caution in mind we may, in addition to those already named, speak of Poland, Bohemia, Hungary, Switzerland, and Burgundy (incorporating varying amounts of the 'Low Countries') as representing fairly definite cultural units. Of these we may recall that Bohemia boasted the earliest university in Eastern Europe (p. 30), Hungary one of the finest libraries; and the fact that Poland at the turn of the century nurtured the young Copernicus gives assurance of a liberal academic culture there. From the heart of Switzerland came the outstanding figure of Paracelsus, but otherwise the country was too closely associated with the neighbouring Italy and 'Germany' for clear distinctions to be made. The same might be said of Burgundy in relation to France; but, though the court of Burgundy in the *quattrocento* had been the centre of artistic activity from which the Italians were glad to learn, of scientific progress there is evidence (e.g. Louvain, p. 102) only a short time before a similar movement could be discerned in France. In Muscovy and the Scandinavian countries it came even later.

Britain, which for most of our discussion can be confined to

England, though in the earlier *quattrocento* more involved in Europe
than it is so far today, was separated by a 'moat' that even in our
own times seems to arouse among the majority of the English a
certain doubt as to whether those living on its far side are properly
to be regarded as belonging to the same *genus homo*. To Scotland
the Renaissance, like the spring, came very late and was of short
duration; the chief cultural influence was French, and the internecine
struggles precipitated at the time of the Reformation by the corrupt
'nobility' and still more corrupt Church precluded any significant
progress in 'science' before very near the end of our period.

The English scene despite, or perhaps because of, the absence of
outstanding figures cannot however be lightly dismissed. As early
as 1439 Duke Humfrey of Gloucester had made a gift of MSS to the
University of Oxford. Subsequent gifts and other MSS still extant
in the Bodleian Library (founded *c*. 1600) and elsewhere are evidence
of a lively concern with Italian humanism. At the turn of the fifteenth century Erasmus acknowledged the pleasant stimulus (despite
the poor food and far from pleasant smells!) he received on his
visits to Cambridge and Oxford; and it was at the latter that John
Colet convinced him of the importance of the teaching of Marsilio
Ficino (p. 152). Nearer to the heart of science was the international
figure of the humanist physician and friend of Henry VIII, Thomas
Linacre. All this disposes of any idea of a cultural vacuum. And the
same might be said of France. It was indeed in Paris that Badius
Ascensius thought it worthwhile to print the first adequate edition
of the works of Nicholas of Cues, edited in scholarly fashion by
Jacques Lefèvre d'Etaples. In 1529 the French king, Francis I,
founded three professorships to offset the theological bias of the
University of Paris; but the interest he sought to promote was that
of literary humanism—it was some time before he would admit to
his own fine library the 'mass products' of the printing press!

Yet in neither England nor France had the growth of humanistic
critical scholarship been accompanied by such a 'scientific' activity
as had taken place in Italy. By the middle of the sixteenth century,
however, there were signs of awakening interest, an account of which
and of the subsequent rapid progress, will be the prior concern of
this chapter. Thereafter, and more briefly, we shall seek in the
nature of their societies possible reasons for the pattern of their
progress.

Towards the end of the *quattrocento* the merchants of Bristol,
no less than those of Nürnberg and Augsburg, were 'looking west'.
Only five years after Columbus had made his first landfall in the
'Indies' the Genoese, John Cabot, having gained from Henry VII
a charter to explore unknown lands beyond the limits laid down by
the Treaty of Tordesillas, set out from Bristol backed by the wealth

of several of its merchants. The prevailing winds prevented John Cabot from penetrating much further south than the inhospitable shores of Nova Scotia; but he had discovered the cod fisheries of the 'New Found Land' and set out again the following year. From that voyage he never returned; and despite a few more Bristolian attempts, including Sebastian's on the 'new hope' of a 'north-west passage' in 1509, the name of England disappears from the annals of geographical enterprise until past mid-century.

While too close a correlation would be hazardous, it is surely significant that in this period of English cosmographical ignorance and apathy not a single printed work on even common arithmetic came from a native press or author until 1543. A laudable if misguided attempt had indeed been made by Cuthbert Tunstall, Bishop of London and later Durham, as early as 1522; laudable, in that the author recognised that even for changing money the academic Boethian *arithmetica* must be replaced by the 'Arabic' style of 'computation' (set forth by the learned merchant, Leonard of Pisa, about 1210!); but misguided, in that the London printer, Richard Pynson, allowed it to be written in Latin and set out in a ponderous style betraying a complete lack of didactic realism. It was never reprinted in England but achieved seven continental editions, perhaps owing to Rabelais' naming it as 'required reading' for the youthful Gargantua. The contrast in 1543 is staggering: the *Ground of Artes* by Robert Recorde, known in twenty-eight editions (the last being late in the seventeenth century), is a *rare* book—there is no copy of any sixteenth-century edition in the remarkably representative collection in the University of Aberdeen; it must have been 'read to pieces'. It is significant also that Recorde was no 'rude mechanical' but a Fellow of All Souls College, Oxford (1531) and Doctor of Medicine, Cambridge (1545), physician to Edward VI (1547), and, though a protestant, to Mary. This remarkable book was followed by a geometry, *The Pathway to Knowledge*, an algebra, *The Whetstone of Witte*, and a highly important book on astronomy, *The Castle of Knowledge*, all published before his death in the prison of the King's Bench in 1558, where he was confined after being found guilty of peculation and maladministration in that grave of many Tudor reputations, Ireland. Involved also in the 'modern' enterprise of mining, and employed at the Bristol mint, this product of the English universities should give us pause before assenting too easily to their reputation for inertia throughout an age of intellectual ferment. In fact an unprejudiced investigation reveals that not only a large proportion of those who were to promote the scientific way of thought in England, but also such brilliant crown servants as Burghley, Walsingham, and the Bacons, all passed through the universities. Of Recorde himself it is true that apart from replacing

the 'rhetorical' symbol 'eq' by the present ' = ' in the 'rule of equation'
he left no mark on mathematics; but in everything he set his hand
to there is evidence of a lively mind, characteristic of the coming
'scientific' approach to the problems that beset the age.

For a decade after Recorde's death there is no outward sign of
any appreciable improvement in contemporary attitudes; this is
confirmed by Henry Billingsley's brief introduction to his (the first)
English translation of Euclid's *Elements*, where he complains of the
'want and lacke of such good authors hitherto in oure Englishe
tounge' despite the fact that 'no artes garnish the mind more than
the mathematicall'. And this he maintained was not due to lack
of demand or native talent among 'gentlemen and others of all
degrees'. The complaint is exactly echoed and amplified in John
Dee's 'very fruitfull preface' that follows; there the wide application
of mathematics is exemplified in a passage suggestive of the modern
cliché 'can we afford not to': 'The difference between the truth
and such erroneous [surveys] would have been able to have found[ed]
(for ever) in each of the two universities an excellent mathematicall
Reader to each allowing yearly 100 markes . . . the famous University
of Paris hath two mathematicall Readers and each 200 French
crownes yearly of the French King's magnificent liberalitie only.'
This *Preface* is one of the most important documents relating to the
growth of science in England.

If Dee's reference to the French king's 'magnificent liberalitie'
refers to the Collège Royale staffed by 'Regius' professors, it involves
a degree of special pleading. The first appointments in 1529 by
Francis I in Greek and Hebrew, followed by Oronce Fine in mathematics in 1532, were regarded as a challenge to the rather scholastic
and too theological university, with which, especially after the
appointment in 1552 of Pierre de la Ramée to the Regius Chair in
Rhetoric, relations became progressively more strained. Fine's
mathematical books are more remarkable for their sumptuous
format and beautiful ornamentation (after his own designs) than
for any outstanding originality. The appreciation of mathematics
in France followed similar lines to those revealed in England, and
with less excuse, since in 1484 Nicolas Chuquet completed a strikingly original and advanced treatise, that had it been printed would
have rivalled Pacioli's (p. 49). Some of it was indeed pirated as
early as about 1538 in a greatly inferior work; but the arithmetic
that seemed to meet most needs was that of Reinerus Gemma
(Antwerp 1540) of Louvain, where Dee found the sort of activity
he hoped to initiate in England. Most notable among his contacts
there was, in addition to Gemma, Gerhard Kaufmann better known
as Mercator.

The value of Recorde's books lay in the fact that though they contain numerous examples of application to practical problems, the methods employed were developed from a sufficient basis of theory to permit of wider application by a reader who took the trouble to study it. There had in fact been 'commercial' arithmetics in France before Recorde's, but they breathed no air outside the counting-house, following the example of the even earlier German 'professional' *Rechenmeister*.

It is not unlikely that in 1552, when Jacques Pelletier published the first 'academic' arithmetic in French, he was at that time mainly concerned to promote the purity and range of the French language, as Luther's Bible had incidentally the German. At any rate he was at pains to justify the inclusion of commercial and similar applications on the ground that such knowledge was 'for the glory of France'. That year—1552—may be regarded as a watershed: his elegant algebra followed two years later, and in 1555 the *Arithmetic* of the by then famous Pierre de la Ramée, who, when recommending the very able Pierre Forcadel for the Regius Chair of Mathematics in 1560, did not hesitate to do so on the ground solely of his mathematical skill, despite his ignorance of Latin and philosophy. The French were coming on.

Pierre de la Ramée himself made no significant advance in mathematics, but by his clear call to widen its field of relevance he hoped to revitalise its study in the universities where the *quadrivium* (p. 20) no longer held the eminent position it had during the High Middle Ages. Even more influential was his challenge to the whole concept of 'higher' education. That any such shock-tactics against the degenerate scholasticism of the day would have been advantageous cannot be denied; but, as almost always happens to innovations, it in turn hardened into a *system* some of whose baleful influences haunt us still. The system came to be known as 'Ramism' (his name being Latinised to Petrus Ramus), and, though far too complex to be given adequate exposition here, must be accorded some degree of characterisation since it powerfully influenced attitudes to scientific knowledge and religion—possibly more so in England than in France.

Ramism may least misleadingly be described in terms of change of emphasis: from Aristotelian dialectic to Ciceronian rhetoric; and from 'analytic' to 'topic'. The aim was primarily to teach the art *bene disserendi*—the 'art of effective discussion'; but note that the Latin *dissero* carries the further meaning of 'set in order'—a fundamental basis of Ramism. And the means adopted to this end was the emphasis on *spatial* (Gk. $\tau o \pi o s$ = place) relationship of ideas. The possibility that this aspect of Ramism may have been promoted by print technology has already been alluded to (p. 64) but its emphasis on *tabular* display of the results of pigeonholing appears to be an exception to McLuhan's insistence on the 'linear' nature of print communication. Such 'topical' displays can express much more complex interrelationships (as in modern flow-sheets and systems-analysis) than could the spoken word; whether Ramus recognised this is another matter.

The influence of Ramism on science was to clarify issues, to convert it from 'wonder' to an organised body of knowledge composed of separate 'subjects' (now replaced by the blessed word 'discipline' in defiance of its original meaning) thereby making it accessible to 'inferior' intellects. An echo of this in Francis Bacon's optimistic forecast will concern us later (p. 159); at this stage it may not be without significance that the 'scientific revolution' is by some historians associated with the English 'puritans'—men who sought to (over?) simplify the subtle distinctions in the relation between God, Man and his Well-Being, and who were demonstrably influenced, notably at the University of Cambridge, by apostles of Ramism.

The slow awakening to new ways of thought in the mathematical sciences was paralleled in Medicine and the 'natural history' related to it. Neither France nor England lacked Greek scholars capable of applying their skill to the production of Latin translations of the medical classics direct from the Greek. First in age and international renown was Thomas Linacre, Fellow of All Souls College, Oxford (where, as well as at Cambridge, he endowed lectures in Medicine), M.D. of Padua, friend of Erasmus and of the famous English 'Grecians' More, Colet, and Grocyn, also of Leoniceno (p. 83) whom he visited at Ferrara. His translations of several of the works of Galen were for the most part first printed in Paris or London, but one was numbered among the very few products of the ephemeral press at Cambridge (1521). In France Jean Ruel produced a new Latin translation of the *Materia Medica* of the Greek Dioscorides, and also, of even greater significance, an 'encyclopedia' of ancient botany based mainly on the Greek of Theophrastos.

Valuable as were these works, neither Linacre nor Ruel displayed critical attitudes towards Medicine itself, as distinct from the language in which it was expressed. Nevertheless in England there occurred a movement of permanent importance: this was the founding (under the patronage of Henry VIII, though the title 'Royal' was not used until 1682) of the College of Physicians of London, and in this movement Linacre played a prominent part. To the same monarch must be accorded the credit for giving effect to a petition by the surgeon, Thomas Vicary, for a union of the rival guilds of barbers and surgeons into one body that demarcated their several fields of practice. From this there later emerged the Royal College of Surgeons of London. Both these Colleges were epoch-making in regulating practice, and laying the basis of two learned professions: an early member, John Banester of the later body, thought that there should be only one, and subsequent history has shown that he may well have been right. Though the first act of incorporation of the physicians stipulated for the provision of executed felons as 'subjects' for anatomies, the members were slow to provide evidence of having taken advantage of it.

Medical studies in France had since the thirteenth century been principally associated with Montpellier—significantly nearer to Italy than Paris—and it was there that the young François Rabelais started lecturing (1531, printed 1532) on the *Aphorisms* of Hippocrates by reference to a Greek text, justifying this apparently pedantic practice by reminding his audience that in Medicine a correct knowledge of the meaning of words might make the difference between life and death.

In establishing even a decade in which could be discerned a more scientific approach to Medicine in England and France greater caution has to be observed than in the case of the mathematical sciences. Progress cannot be so confidently documented by reference to the printed sources. Though the restrictions of anatomies have been generally exaggerated there *were* restrictions, and the activity of the artists of the High Renaissance (p. 45) should give us pause before inferring too much from the *absence* of evidence. We know for example that the basis of Vesalius's epoch-making studies in anatomy was the skeleton of a felon he cut down from the gallows— an action it would have been imprudent to advertise even to ensure priority of publication! There is similar indirect evidence (though obviously for different reasons) of the use of mounted specimens (*herbaria*) of plants, and of the teaching of medical botany both by lecture and demonstration in Italy and Montpellier before the appearance of any *critical* printed works in those countries respectively. In England William Turner, one of the greatest naturalists of the sixteenth century, wrote in 1548 that he had refrained from publishing a herbal already written in Latin until he might 'declare to the greate honoure of our countre what numbre of sovereine and strang herbes were in Englande that were not in other nations'. Incidentally this same 'nationalistic' note was sounded in the more literary *Hortus Gallicus* ('French Garden') by the French scholar and antiquarian Symphorien Champier, and also in the much more critical and 'modern' work by the German, Jerome Bock.

When all this has been said it nevertheless remains true that, though at least a decade before the general awakening in mathematical thought in France and England evidence is forthcoming of a fairly widespread academic interest in critical natural history in those countries, no convincing case could be made out for regarding any of their natives as pioneers in the field. William Turner has perhaps the best claim, but even he insisted that so ignorant of herbals were the physicians while he was at Cambridge that to gain knowledge of plants he was compelled to travel in *Italy, Germany*, and *Switzerland*. It was in Bologna (Italy) that he probably graduated M.D. and certainly studied under Luca Ghini, the first teacher in Europe to be appointed (1544) expressly in medical botany. From Cologne came Turner's remarkable book on birds with evidence of dissection for the purpose of checking the descriptions of Pliny; and in Zürich he became the friend of Conrad Gesner, author of the first comprehensive encyclopedia of zoology for which Turner himself supplied local information. Nor can there be any reasonable doubt that a solid basis for scientific botany was laid by the Germans, Otto Brunfels, Leonhart Fuchs (p. 94) and Jerome Bock—all before 1543.

A lack of interest in human anatomy in France and England is even more striking. Before 1543 when the appearance of Vesalius' work marked the beginning of a new epoch there existed only one not very convincing work by Charles Estienne in France and one by David Edwards in England. The latter, only a brief compilation based on the medieval 'classic' by Mondino, is saved from insignificance by an indication that the author had, probably in Oxford, carried out some anatomical investigations. Neither of these is on a par with those written by three Italians and one German in the same period. But if Brussels and Louvain could in any sense be regarded as in 'France' then it would appear that by far the greatest contribution came from that quarter. It has however to be noted that Andreas Vesalius *Bruxellensis* had the chance of demonstrating on the cadaver for the lectures of Jacques Dubois (better known as 'Sylvius') in Paris, where he was also greatly encouraged by the German Iohann Guinter of Andernach, and perhaps stimulated by the rivalry of the young Catalan, Miguel Servet (later burnt by Calvin for mixing physiology with theology); and finally that the greater part of the necessary research was done by Vesalius after his appointment in 1537 as Professor of Anatomy and Surgery (note the combination) in Padua. The book itself first saw the light in Basel, where the printer, Oporinus, must have risked an enormous amount of capital. To this extent was the anatomical revolution an international affair. The share of Louvain lay in the classical and Hebrew scholarship that the young Vesalius acquired in its (the first) 'trilingual' university. From Guinter he acquired only his medical scholarship; the German remained satisfied to produce the best possible Latin *text* of Galen's foundation work *On Anatomical Procedures*—an indispensable task at that time. To Sylvius Vesalius perhaps owed more than he was later willing to admit; but the master's response to his pupil's revolutionary work was a scurrilous attack on the 'madman' who had written it. Not in Paris, then, but probably only in Italy were conditions propitious for launching a work that Charles Singer described as 'suddenly, essentially, and brilliantly modern'. Modern it certainly was, not least in its sumptuous format, superb illustrations, and (in the dedication to the Emperor Charles V to whose household Vesalius was later attached) in the unabashed and also incoherent claim to its unique excellence. But since scholars have been less prone to take Vesalius (then only in his twenty-eighth year) at his own valuation the 'suddenness' of the modernity becomes somewhat less convincing.

The title of the book is *De Humani Corporis Fabrica*; the 'working' of the human body is perhaps the best translation, since *Fabrica* connotes more than mere 'structure'. Its 'modernity' on the constructive level consists in shifting the emphasis from the *dead* to the *living* body, involving a knowledge of the supporting skeleton before the demonstration of the action of the musculature; in its evident (and too much reiterated) basis in the first-hand experience of the author; in the use of an array of *specialised* instruments for operations ranging from the division of a bone to the tracing of a minute branch of a nerve; and, above all, superbly beautiful 'action' pictures. The still common assertion that these were the work of his fellow countryman, Jan Stevan van Calcar, a pupil of Titian, can hardly be maintained when they are compared with those signed by Jan Stevan among the six diagrams prepared separately a few years earlier; they are in any case of uneven merit. Who the artists of the *Fabrica* were remains a strangely unsolved mystery.

In each one of these 'innovations' Vesalius had been anticipated; but by no one (except Leonardo) had he been forestalled in respect of them all. With all his faults his position as a supreme master of descriptive science (without which the obstacle of the ingenious but too comprehensive physiology of Galen could never have been removed) remains unassailed. Far more immediately influential than Copernicus, by his single-minded dedication during less than a decade he changed the face of medical education by removing it from a predominantly literary setting. But it took a great deal longer to break the thraldom of the traditional physiology: nearly ten years after the *Fabrica* had seen the light in Basel the greatest of the printers of that city, Froben and Bischoff, brought out a magnificent four-volume edition of the works of Galen.

This major advance towards an empirical basis for biological thought has been discussed in this context to emphasise the fact that though the advanced climate of opinion of Italy was still a decisive factor, it was not an Italian who played the decisive part.

If we may now assume that the time-lag in scientific activity in France and England has been sufficiently elucidated, we may attempt to assess the progress during the second half of the century. Though the Treaty of Cateau-Cambrésis put an end to the wasteful folly of the French 'adventures' in Italy, the infiltration of emissaries from the theocratic 'state' founded by the Frenchman, Jean Calvin, in Geneva, no less than severe internal tensions, went a long way towards guaranteeing that within three years the French would justify the clash of latent political ambitions by a 'holy war'. The hideous orgies of massacre and destruction that rent the country in an almost unbroken sequence until near the end of the century were hardly conducive to solid progress in the sciences; they are bitterly lamented in the preface of more than one scientific book of the period. In England a similar disaster could not have been ruled out: attempts on Elizabeth's life were probably more numerous than those on Catherine de' Medici's. But the contrast between the former's political intuition—wisdom might be too strong a word—and the narrow ambitions, furthered by ill-judged opportunism, of the latter was probably one of the major reasons for England's escape. At all events the intelligently directed scientific activity in England, though not productive of any spectacular achievement until the turn of the century, was sufficiently more notable than that of France to justify our once more taking the former as the paradigm.

Billingsley's complaint (p. 102) regarding the absence of mathematical books in English was published in 1570; within the same decade appeared three books by Leonard Digges, one of which had been supplemented and two actually completed by his son Thomas. The *Pantometrica* (1571) and *Stratiotikos* (1579) appeared for the first time; the *Prognostication everlasting* . . . had already (1576) gone through four editions before Thomas added a chart and text that made of it one of the most important documents in the science of the Renaissance. These works were originally conceived as technical handbooks on surveying, military science, and meteorology with a distinctly astrological flavour; in passing through Thomas's hands they received the stamp of a highly skilled geometer and astronomer more advanced in cosmological thought than is to be found in any other contemporary evidence. In *Pantometrica* Thomas emphasises that there were living witnesses to his father's remarkable achievements with 'proportionall ['perspective' in title] glasses'. There is now no reasonable doubt that this refers to some kind of telescope about thirty-five years before the invention of the refracting telescope registered in Holland in 1608.

There were many more books relating to the 'mathematical arts' published in England before the end of the century; all were concerned with 'arts'—navigation, cartography, dialling, terrestrial magnetism, surveying, gunnery—rather than with mathematics itself, in which Thomas alone had shown himself to be an original thinker. The importance of these works lies, then, in the evidence they provide for the realisation that mathematics is a necessary basis for the *improvement* of these 'arts'. They reveal also the presence of men sufficiently versed in mathematics to be able to undertake the efficient translation of works in other languages, such as the well-known *Arte de Navegar* by Martin Cortes, translated from the original Spanish (1551) by Richard Eden in 1561. Though the incursion of men of 'book-learning' into the field of navigation was resented by some of the 'practical' men, it is now generally agreed that the remarkable voyages of, for instance, Richard Chancellor and John Davis would hardly have been possible had they not been willing to submit themselves to 'land-based' teachers such as Digges and John Dee.

Since most of these works are relevant to the spread of 'quantification' in all its aspects some of them will be accorded more detailed notice in the chapter relating to that subject; but recent examination of the navigational and cartographical manuscripts of Thomas Harriot has revealed the fact that before the end of the century this remarkable man was a master of purely mathematical methods far surpassing anything seen in England at that time. His mathematical genius stands out even more clearly in his algebra, of which a sufficiently startling version was published only some years after his death. Unfortunately the confused state of his MSS does not permit of any decision as to how far the printed work represents Harriot's own original methods. At least one of his editors, Nathanial Torporley, had by then had the advantage of studying the collected works of François Viète, almost the sole, and certainly the greatest, glory of French mathematical science up to that time. But there is little doubt of Harriot's superiority to Viète in respect of algebraical notation, and there is even less with regard to the originality of his optical discoveries. His independent telescopic observations may not have been so comprehensive as those of Galileo but he had a far deeper understanding of the theory of refraction than had the latter. Though it seems unlikely that Harriot's mathematical genius was in need of any extraneous motivation, it is a fact that his contemporary European reputation was based on his scientific survey in North America, parts of which were published as *A briefe and true report of the new found land of Virginia* in 1588, when he was still only twenty-eight. The critical assessment of the characteristics and resources of the country (not identical with the modern state of that name) is almost the only original English contribution to the advance of natural history: the earlier achievement of William Turner (p. 105) was not equalled until much later—unless John Caius's comparative study of English dogs can be so rated. In this department France and the bordering Low Countries excelled.

The centre from which most of this activity radiated was Montpellier where before the civil wars a progressive school of medical
botany had already been established. Outstanding among the
teachers was Guillaume Rondelet who, though he published nothing
on plants, could claim among his pupils most of the workers whose
names are familiar to us today: Felix Plater, under whom the
medical school of Basel achieved a fame comparable to that of
Padua; Jean Bauhin, whose son Gaspar was the first scholar to be
appointed as Professor of Anatomy and Botany (also at Basel);
Jacques d'Alechamps, author of a great compendium of sixteenth-
century botany; Matthias de l'Obel, perhaps the first to construct
a systematic classification of plants on supposed natural relationships; Charles de l'Ecluse, a polymath best known to botanists for
his translations into Latin (one undertaken while waiting in the
Thames for a favourable wind) of the important works on exotic
plants by the Spaniards, Cristobal Acosta and Nicolas Monardes,
and the Portuguese, Garcia de Orta. De l'Obel, de l'Ecluse, and
their older contemporary, Rembert Dodoens, were natives of Lille,
Arras, and Malines respectively. They were bound together in such
intimate mutual affection and regard that it is impossible to distinguish comprehensively between their respective contributions to
knowledge. The extensive market for these books may be gauged by
the fact that Christophe Plantin of Antwerp found it worthwhile to
bring out a collection of the pictures from these books without text;
this unusual format might claim to be the forerunner of the coffee-
table style! Both de l'Obel and de l'Ecluse spent much of their
lives in England laying out gardens and contributing new identifi-
cations to the English flora; for comparison with the descriptions
contained in the books he was translating and editing the latter made
use of specimens brought from America by Francis Drake. Rondelet's published work was on fishes; in this he provided a great mass
of information on a large number of species. His contemporary,
Pierre Belon, wrote more critically on the same subject, but the work
on the Woodpecker's larynx by the still too little appreciated
Volcher Coiter of Groningen came nearest to the achievement of
comparative anatomy, so important for Harvey's work on the
motion of the heart in animals.

Of the growing importance of military surgery a hint has been
given in the previous chapter; of the more enlightened 'profession-
alism' of surgery in England something has been said in the present
chapter (p. 104). It now remains to mention the change of method
and perhaps still more important the changes in attitude brought
about by the immortal Ambroise Paré. The story of his accidental
discovery of a new technique for the dressing of wounds and his
consequent rejection of the traditional routine of applying hot oil

is well known: he ran out of oil and tried a cool 'digestive of eggs, oil of roses, and turpentine'. The following morning those treated in the customary manner were feverish, swollen, and in pain; those by the 'digestive' were 'fairly comfortable' as the saying is. Thus was established in his mind the conviction that surgery must be learnt by practice rather than from books; hence, though ignorant of Latin, he might succeed where the learned surgeons of the Collège de St Colme failed. And succeed he did, aided by a ready pen and a telling wit, against bitter opposition, in becoming surgeon not only to four successive kings of France, but, as Dr Douglas Guthrie has remarked and as Paré's great book also revealed, 'one of the master surgeons of history'. His success depended not a little on his intuition—expressed in his oft-repeated saying *Je le pansai . . . Dieu le guerit* ('I dressed him . . . God cured him')—that there was not a great deal that medical 'science' could do to effect a cure except to provide the most favourable conditions for the *vis medicatrix naturae* ('the healing power of nature')—a lesson that the physicians only learned a century later from the books and teaching of Thomas Sydenham.

It would be misleading to make comparison in respect of France and England only on the basis of books printed before the end of the century. In 1600 appeared William Gilbert's book with the revealing title: *On the Magnet, magnetic bodies, and the great magnet the Earth; a new physiology demonstrated by numerous arguments and experiments* (original in Latin); there is no doubt that the greater part of the work had been completed several years earlier. It would be generally agreed that it was the first work in which the 'new' experimental method of *enquiry*, as distinct from exemplification or the attainment of some technical aim, is dominant. Its larger consideration must be deferred (p. 67); it is mentioned here since it has strong claims to be the crowning achievement of the remarkably rapid growth of a 'scientific' approach to the 'arts' in the last decades of Elizabethan England.

Of more limited extent at the time, but fraught with comparable significance for the experimental method of enquiry, was the concern with mining and metallurgy signalised by the granting of royal charters to the Company of Mines Royal and the Company of Mineral and Battery Works. The great treatise (*De Re Metallica*) on mining and metallurgy (and related industries, such as glass manufacture) by the German, Georgius Agricola, had been available since 1556, and the *Pirotechnia* of Vanoccio Biringuccio (to which Agricola owed a good deal) from 1540. The latter appeared in French not later than 1556; but no work on the subject was printed in English until the following century, by which time the more important treatise by Lazarus Ercker was available. The English enterprises were thus dependent on the knowledge and skill of imported German workers; but there is manuscript evidence of intelligent appraisal by one of the native 'managers' of some of the more puzzling chemical issues involved (p. 137).

From this sketch, embodying a whole century of scientific activity in France and England, the reader must be warned against drawing any detailed conclusions: the absence of detailed reference to work in other countries (as later chapters will show) does not imply lack of achievement of the highest importance. Evidence has been put forward in support of the view that the influence of science on the societies of the great monarchies was delayed by about half a century; that during the second half of the century it was, except in the biological field, more marked in England than in France; and that in the former country it had at least the appearance of being the consequence of political and social demands, and was most successful in those fields where the backing of state or mercantile power was essential to effective progress. The rapid growth of (mainly navigational) scientific instrument production in London points in the same direction. In 1580 Robert Norman in his book *The New Attractive* showed that some of the men engaged in this trade resembled in some respects the artist-craftsmen of the Italian *quattrocento*. A recent addition to the British Museum collection of clocks is a reminder that a similar influence was not wholly absent in Paris as early as 1544 when Fleurent Valleran and six associates succeeded in extracting from Francis I a charter establishing the first Guild of Clockmakers in Europe: horology thus gained recognition as a specialised 'art', whose subsequent importance for science needs no emphasis.

It is reasonable to ask ourselves whether there was any factor common to France and England that could account for the very marked underdevelopment of interest in natural knowledge and even of what the medieval English friar, Roger Bacon, had called its 'Key', mathematics. There was indeed one factor common to their political development. This was the tendency to enlarge the extent of royal jurisdiction, and by changing the relation between the crown and its subjects to concentrate power at the centre. It has become customary to apply the term 'nation-state' to the consequent political unit. The novelty of this 'abstraction', as Mattingly appropriately called it, has been obscured by the components of the term used to name it. The medieval world was familiar with both 'nations' and 'states': the former were an important factor in university governance, but these followed boundaries quite different from the 'states' that might bear the same name. The novelty lies in the gradual merging of the two components into one entity. To what extent was this process peculiar to England and France?

In Italy during the first half of the *quattrocento* the rival territorial claims and shifting alliances of the five dominant states—Milan, Florence, Venice, Naples, and the Papacy—precluded the emergence of any *Italian* state. At one time (p. 25) it looked as if Milan might absorb the remainder; but only a tenuous equilibrium was established. During this period of about forty years the peninsula came as near as it ever had done to the reign of peace; and these years were also those of the flowering of the Italian Renaissance. But though this reached its peak in the first decade of the *cinquecento* any hope of political unity was destroyed in 1494 when Lodovico Sforza invited Charles VIII of France to help to establish the long-sought Milanese hegemony.

In the same forty years England emerged from the last serious threat to the royal power and the King of France finally broke the power of his feudatory Duchy of Burgundy. Where France and England differed radically from Italy was that in both there had been for at least two centuries a royal power to challenge. As the *quattro-cento* closed, Henry VII was developing a professionally manned machinery of government of great significance for the next century. Though the central power of France had been much strengthened, the successive invasions of Italy revealed a hangover of the feudal concept of royal chivalry and the lure of hereditary right to distant territories unconnected with national advantage: wealth and energy that might have gone into cultural advance were thus drained away for half a century.

The only other territory comparable to England and France in respect of centralisation of royal power was the Iberian Peninsula after the union of the crowns of Castile and Aragon. A case might indeed be made out for regarding the nations of the Iberian Peninsula as intermediate between Germany and Italy on the one hand and France and England on the other. For while there was undoubtedly a manifestation of scientific enterprise first in Portugal, somewhat later in Spain, and in both about half a century before anything comparable happened in France or England, yet it was of a kind—geographical exploration backed by systematic application of astronomy and mathematics—that could hardly have been undertaken except in the context of material resources and political and military power associated with centralised government. It may be significant that the scientific 'fall-out' resulting from this enterprise, in the way of a vast addition of natural phenomena to be classified and related to existing knowledge, was little developed until the other nation-states were sufficiently on the road to scientific understanding to be able to share in the burden of study.

It is unlikely that the early concentration of royal power in France and England was in itself a significant factor in the relatively

slow response to the Renaissance impulse. In the *trecento* Paris and
Oxford had been pioneers in a kind of 'mathematical physics' that
the Italians cultivated only a century later. More important, perhaps,
was the shift in the balance of power between secular and ecclesiastical
authority. When Henry VIII declared himself Head of the Church
of England he was doing little more than make patent a claim implicit
from the time of the fourteenth-century Statutes of Praemunire and
Provisors that prescribed dire penalties on any English subject
appealing to Rome against a decision of the royal courts. The
Gallican Church came later and was somewhat less drastic in its
assertion of independence.

The 'Reformation' in England went in fact far beyond Henry's
intention: the corruption of many monastic houses, the reversal of
the medieval distinction between a learned priesthood and a relatively illiterate laity, and the general restlessness against authority
of every kind—these social factors had more to do with the triumph
of protestantism than had Henry's connubial irregularities. Though
there is no evidence for the belief that the consequent radical change
in the national church created an atmosphere in which 'science'
could flourish untrammelled by the restriction of thought imposed
by the Church of Rome (pp. 122 and 127), it would be equally mis-
taken to dismiss the Reformation as a 'squabble of monks'. It is
hardly conceivable that Luther's emphasis on the 'priesthood of all
believers' would not have been interpreted by some more enterprising
spirits as a licence to question the authoritarian imposition of a
dogmatic natural philosophy. What they may not have realised,
but many have since learnt to their cost, is that the authority had
merely been transferred to the literal text of the Bible. A social factor
which, especially on a long-term reckoning, may have been the most
important condition for ultimate scientific progress was what has
been described as the 'rise of the gentry'—or at least of some of the
gentry. The rapid recovery of national prosperity under Henry VII
was the consequence not only of the replacement of anarchy between
warring magnates by firm central government but also of the elevation of men of the 'middle sort' to high administrative office. The
term 'middle sort' is used as a warning against any of that *pseudo*-
precision associated with the relatively modern 'middle class'.
No greater precision can be given to the 'middle sort' than to regard
them as those who, though bound to no master except the King,
made no claim to patent of nobility. It is sometimes said that these
were mainly the town-based ('bourgeois') merchants; but there were
always exceptions: in fourteenth-century England—and even before—
the names of nobles who had possessed the same estates for centuries
can be detected in the annals of entrepreneurial activity of various
kinds. To complicate matters still further there were families, like

the oft-cited Pastons, who, though otherwise qualifying as of the 'middle sort', were landed 'gentry'. Such families may indeed have had a special significance for our enquiry since the records show that it was largely with their sons that the rising Inns of Court were filled. The critical study of law, though not in itself especially conducive to scientific speculation, called for a great expansion in 'higher education' for the laity. In a phrase applied to Francis Bacon it 'called the wits together'; and it effected a shift of emphasis from knightly to literary accomplishment. In Elizabeth's reign the universities of Oxford and Cambridge might have been a good deal more moribund than they actually appear to have been (p. 100) had their curricula not been somewhat reformed, at least in respect of the kind of questions debated. However that may be, there is no question that by then there was a marked increase in the number of 'gentry' ready and eager to take seriously the new possibilities of natural knowledge. Here England differed from France: no French bishop wrote a text-book on arithmetic as an aid to money changing; Billingsley's reference to 'talent and demand among gentlemen' (p. 102) would have been regarded as in doubtful taste. It is true that nearly a century later the Hon. Robert Boyle was still bewailing the reluctance of gentlemen to soil their hands with charcoal; but a few years after that Thomas Sprat could justly claim that the Royal Society was largely composed of 'gentlemen free and unconfined'.

The inconclusive note on which this chapter ends is intentional: to the author there is as yet no conclusion in sight. What has been attempted is to demonstrate a marked difference of phasing in the emergent nation-states as compared with that of the more loosely articulated pioneers of Renaissance science and to isolate such socio-political factors as seem relevant to this difference; not forgetting less patent differences between the nation-states themselves. To single out any of these factors as decisive would, in the present state of our knowledge, be to substitute plausibility for evidence.

One event in England is apt to pass unnoticed, since its effects were delayed for about thirty years: the bequest by Sir Thomas Gresham, Elizabeth's financial adviser, of his London residence and (subject to his widow's life-interest) his mercantile property within the City of London for the foundation and endowment of an independent college of resident professors in mathematics, astronomy, medicine and other 'arts'. What might have been the outcome had his widow been more 'accommodating' is an interesting conjecture; what actually happened will be referred to in Chapter 12.

9

THE COPERNICAN REVOLUTION

In the spring of the year (1543) that Vesalius' *Fabrica* was published in Basel, Mikolaj Kopernik of Torùn in Poland lay dying in Frauenburg. There, as a secular canon of the cathedral, he had spent the last forty years of a life divided between the political service of his country and the enrichment of human learning in many fields (p. 136). A pious tradition relates that before his senses faded he was able to lay his hands on the first copy, fresh from Nürnberg, of the book embodying the fruits of his life's devotion to astronomy— *De Revolutionibus Orbium Coelestium*. The word 'orbes' is ambiguous in his writings—sometimes it stands for what we should call 'spheres'; in other places 'orbits' would be more appropriate; the point is not unimportant. The book had not come unheralded. More than a score of years earlier he wrote out a sketch (known to scholars as the *Commentariolus*) setting forth the hypotheses of the system he ultimately adopted, but without detailed demonstrations. This was the fruit of his youthful study, first at the University of Cracow, later and for a much longer period in Italy: at Bologna where, assisting the astronomer Domenico of Novara, his observations cast doubt on the adequacy of the contemporary theory of the Moon's motion; at Padua, where Luca Gaurico (p. 34) and Fracastoro (p. 95) were probably his colleagues; and finally at Ferrara, where he graduated as Doctor of Canon Law. The *Commentariolus* was never printed; none of the copies circulated to friends and other astronomers survives; only three transcripts (at Vienna, Stockholm, and Aberdeen), made soon after the author's death, are known. Of any detailed knowledge of its reception we have none, despite the author's association with leading Italian scholars at Ferrara. Nothing more was heard of the project until 1540, when a *First*

Sketch (*Narratio prima*) of the coming *De Revolutionibus* was issued by his ardent young disciple and co-worker, Joachim Rheticus. This kite not having been shot down by any ecclesiastical heresy-hunter, Rheticus persuaded his 'teacher' to allow him to take the MS of the *De Revolutionibus* to Germany where it was printed by Ioannes Petreius of Nürnberg.

The circumstances of the beginning of the 'Copernican Revolution' are still not as widely known as is necessary for a just appreciation of its own significance and of the contemporary reactions. Examination of the book itself reveals the fact that it bore a recommendation from a Cardinal and was dedicated to Pope Paul III. On the other hand Luther, who had a preview of the MS borne first to Wittenberg by his follower, Rheticus, described the author as an 'ass' (*Narr*) who merely wanted to gain notoriety by turning things upside down. Over seventy years passed without any marked signs of other ecclesiastical displeasure; but in 1616 it was placed on the Index of books not to be read by the laity until 'corrected'. Although hailed by some modern writers as marking the 'beginning of modern science', the fundamental theory expounded by Copernicus was rejected by Tycho Brahe, the greatest astronomer of the sixteenth century, as 'absurd'. Praised for its mathematical ingenuity by dozens of reasonably competent critics in the decades following its publication it received as a serious contribution to astronomical theory no more than a casual word of welcome (and that in a book on astrology) until 1573 when Thomas Digges hailed it as a highly acceptable advance on the traditional theory. By no one was it used as a basis for the reform of astronomical science until near the end of the century, when Iohannes Kepler adopted it as the fundamental assumption in an otherwise fantastic theory (p. 144).

What, then, was the nature of this work which provoked so many contradictory reactions? So far it has been possible in this book to avoid any detailed consideration of the technical aspects of the 'science' whose relation to Renaissance society is the main concern. In the present chapter this can no longer be avoided, since the nature of the reactions—or even of the absence of any reaction, as in Fernel's rhetorical flourish quoted at the head of the last chapter—arose out of the technical issues at stake. So far as the customary transactions of everyday life are concerned Copernicus might never have been born: it would be absurd to insist on children saying 'the Earth is now turning towards the east' instead of 'the Sun is setting in the west'. In the words of one of the most critical minds of a later century, George Berkeley, 'in such matters we should think with the learned but speak with the vulgar'. Even a highly technical reference-work like the *Nautical Almanac* still represents the movements of the heavenly bodies as seen from a *stationary* Earth; for that is how they *are* seen. It is only when we are concerned, as Copernicus and Tycho Brahe were, with some way of avoiding the discrepancies between forecast and observation that assumptions going beyond immediate visual experience have to be made. The Babylonian astronomers of the fourth century B.C. devised a purely 'operational calculus' that allowed remarkably good predictions to be made without any such assumptions. The Greeks, whose world-view Western society inherited, in almost total ignorance of the advanced mathematical astronomy of the Middle East, took a different view of the matter. Their insatiable curiosity and love of argument, combined with a powerful visual imagination, compelled them to create a κοσμος (ordered system) out of the χαος (confusion) of human experience, to which the starry heavens were the only exception—but not quite: the planets (among which were included the Sun and Moon) displayed a degree of disorderliness that could not be tolerated; and so the trouble began.

Shorn of those technicalities not pertinent to our purpose the outcome may be summarised as follows: The Earth (*terra*) was motionless at the centre of the world (*mundus*) and the scene of ceaseless interaction between the four elements—earth, water, air, fire. Beyond 'fire' was the 'upper air' (ἀιθγρ=ether) and the 'sphere' of the Moon. This 'sphere' was succeeded by those of Venus, Mercury, the Sun, Mars, Jupiter and Saturn. The 'spheres' were incorruptible and concentric with the Earth about which they rotated each with a characteristic *uniform* speed. The whole was encompassed by the 'eighth sphere' of the fixed stars. The events in the 'sublunary sphere' were the concern of 'physics', in the 'celestial' that of 'astronomy'.

Of course resort had to be made to various geometrical devices to account for the 'phenomena' (appearances), namely, that as seen from the Earth the motions of the planets were very far from uniform. By the time (third century B.C. to second century A.D.) of the Golden Age of Greek (really Hellenistic) astronomy centred on Alexandria, a tradition had been established sharply distinguishing *astronomy* from *physics*; the former (really applied mathematics, hence the custom of calling court astronomers *mathematici*) was concerned solely with the prediction of 'appearances' from 'appearances'; the latter sought the cause of why these 'are so and could not be otherwise' (p. 92).

The stage was thus set for the final synthesis in Ptolemy's *Almagest* (*c*. A.D. 150). Accepting—a fact not always made clear—Aristotelian 'physics', and selecting the most appropriate devices from the works of former astronomers, Ptolemy devised a kinematical-geometrical model of great complexity by means of which remarkably good predictions were made possible, and which was not effectively changed before Copernicus—and only partially by him. Seeing that the 'spheres' were redundant for *astronomy*, Ptolemy (as had many before him) simply ignored them. On the other hand he accepted as fundamental the essentially *physical* assumptions of a central stationary Earth and uniform circular motion for everything else 'above the region of the elements'. To 'save the phenomena' ($\sigma\omega\zeta\epsilon\iota\nu\ \tau\alpha\ \phi\alpha\iota\nu\omega\mu\epsilon\nu\alpha = $ *servare apparentias*) he had to make the centre, to which the circular motions were referred, *ex*centric to the Earth; and to account for the 'retrogradations' of the planets he represented them as points moving on circles each of whose centres in turn moved on a separate circle (different for each planet) about the excentric; this sort of motion is known as 'epicyclic', and is well known in gears and other mechanisms. The balancing of the various radii and circular velocities was of great complexity, but fortunately does not concern us. Unfortunately, there is one other element in Ptolemy's model that does; its omission from some simplified accounts precludes a proper understanding of the modifications proposed by Copernicus: this is the so-called equant-point—distinct from the excentric—about which the several angular *velocities*—but not the actual paths—of the celestial bodies remain constant. The importance of this is that it follows from an essentially *physical* assumption, that of uniform circular velocity; but the equant-point itself has no physical significance in the heavens. The excentric admittedly introduces an element of arbitrariness; but its position is determined with respect to the Earth. The equant-point carries the arbitrariness a stage further. Let us now see what changes Copernicus made in this model.

The 'revolution' introduced by Copernicus was of course to set the Earth in motion, and to place the Sun at the centre of the world. But though constituting a 'revolution' in relation to contemporary attitudes, neither was 'revolutionary' in the sense of being an unheard-of novelty. Copernicus indeed went out of his way to emphasise that both suggestions probably had sound classical parentage though were never generally admitted by the Greeks; he was in fact unaware that the Greek, Aristarchos of Samos, had categorically made the same assumptions and been rebuked for doing so. For us it is perhaps more important that the question had been raised in the High Middle Ages 'whether the Earth remains quiescent in the centre of the world' and answered in the affirmative, not by ecclesiastical direction but because there seemed to be no adequate reason for supposing it to be in motion. In the fifteenth century Nicholas of Cues, who had studied in the same university (Padua) as did Nicholas Copernicus about half a century later, made the much more revolutionary claim that not only the Earth but every object in the world must be in motion and that the unbounded

extent of the latter implied that it was *without* centre. This marvellous
speculative leap was supported by ingenious mathematical analogies;
but being no great astronomer (nor for that matter a very *competent*
mathematician) Nicholas' views seem to have been without wide
influence. A well-known contemporary of Copernicus, Celio
Calcagnini, bewailed the fact that he had become acquainted with
the speculations of the Cusan only after his own similar ones had
been more crudely adumbrated; rather oddly he does not mention
Copernicus at all. A century was to pass before a man of similar
imaginative power, Giordano Bruno, was to develop the suggestions
of 'the divine Cusanus' (*il divino Cusano*) into a system of startlingly
modern appearance (p. 127).

Where Copernicus was ahead of the Greeks was in his recognition
of the *immensity* of the World, and consequently that of the two
'absurdities'—the daily rotation of the 'sluggish' Earth and that of
the incomparably greater 'eighth sphere' (p. 118)—it was more
philosophical to adopt the less absurd. Thus, one of the complexities
of the Ptolemaic system was removed; by giving the Earth an annual
motion round the Sun he was able (like Aristarchos) to get rid of
another. Nevertheless two serious difficulties remained. If the model
was to be of any use for the prediction of positions within the limit
of 10′ of arc (the greatest accuracy then obtainable) he found that
the centre to which the movements of the planets had to be referred
was not the Sun but the centre of the Earth's (excentric) orbit—a
fact not revealed on the famous plan of the system placed near the
beginning of his book and always reproduced in illustration. This
was bad enough; but worse was the impossibility of merely ignoring
the physically meaningless equant-point—neither the Sun, nor the
centre of the Earth's orbit. Copernicus found (as unknown to him
had at least one Arab astronomer) that the introduction of a secondary epicycle produced almost the same result as the equant. He
regarded this as his greatest achievement; and in respect of its
mathematical ingenuity it has been accepted as such by Dr Otto
Neugebauer, whose recent reassessment has been extensively drawn
on in this chapter. Nevertheless, since this new device is effective
only for orbits of small eccentricity, its adoption was only post-
poning the evil day; with the more accurate data bequeathed to him
by Tycho Brahe (p. 127) Kepler found that the equant had to be
restored even for the Earth's motion.

As Copernicus himself was only too well aware, his system flew
in the face of both common sense and the universally accepted
cosmology. In addition, though simpler to visualise at a superficial
level, it is more difficult to apply in the task for which contemporaries
saw its greatest benefit—the calculation of tables of planetary positions. Moreover the so-called Prutenic Tables, calculated by Erasmus

Reinhold, though superior to the Alphonsine that had served astronomers since the thirteenth century, had too slender a basis in observation to have more than a comparatively short effective life. On what grounds then can Copernicus' achievement be claimed as marking the 'beginning' of modern science? Precisely for the reason that it *did* reject common sense as an ultimate court of appeal. The 'structure of a scientific revolution', as Professor T. S. Kuhn has persuasively urged, is 'the setting up of a new paradigm'. Since this commonly involves the rejection of immediate experience, as Copernicus' rejection of any real motion of the Sun did, it is a rejection of the empiricism that has commonly been regarded as the fundamental basis of science. Such a general decision therefore is not properly a 'scientific' but a philosophical one.

In an age when the established philosophy was closely bound up with ecclesiastical power, and the price of such a decision might be —though it is doubtful whether by itself it ever was—torture or even death, Copernicus' decision was a bold one. Unfortunately for the progress of science it was not bold enough: it rejected only the crudest elements of the Aristotelian geocentric philosophy. Copernicus never seems to have asked himself whether the dogma of uniform circular motion about a significant centre was any more than a provisional (though in its day ingenious and fruitful) hypothesis; hence the fuss about the excentric and equant points. In fact, so far from questioning its validity, he enlarged on it in a hymn to the Sun that stands very oddly in a place where he has just appealed from 'common sense' to 'mathematicians':

In the middle of all sits the Sun enthroned. In this most beautiful temple could we place this luminary in any better position from which he can illuminate the whole at once? He is rightly called the Lamp, the Mind, the Ruler, of the Universe; Hermes Trismegistus names him the Visible God, Sophocles' Electra calls him the All-Seeing. So the Sun sits upon a royal throne ruling his children the planets which circle round him. . . . Meanwhile the Earth conceives by the Sun and becomes pregnant with an annual rebirth. . . . So we find underlying this ordination an admirable symmetry in the Universe, and a clear bond of harmony in the motion and magnitude of the orbs, such as can be discovered in no other wise.

In this astounding passage Copernicus replaces the 'established' philosophy by one at that time believed to be much older (p. 153). How much better fitted it might be as a cosmological stage on which the play of modern science was to be acted is open to question, but it is now beyond dispute that many of the chief actors in the play were strongly influenced by it.

Otherwise it would be tempting to dismiss the passage as no more than a rhetorical outburst; such a view seemed to be supported by an anonymous Foreword in which it is emphasised that the whole book was to be regarded as a mathematical exercise to demonstrate how the 'phenomena' *could* be 'saved' without raising the question as to whether the 'hypotheses' were actually 'true'. The presence of this *caveat*—at that time presumed by most readers to have been written by Copernicus—or his disciple Rheticus—may account for the fact that no ecclesiastical voice was raised against even the teaching of the system until much later; if Galileo had been content to interpret it in this way he would have remained an ornament to the Faith instead of being branded as a dangerous heretic.

It is more than probable that close associates of Copernicus knew that the *caveat* had been inserted without his permission—its inconsistency with the general tone of the book is obvious. The writer's identification as Andreas Osiander, a Lutheran minister to whom Rheticus had to leave the final arrangements for printing, was first made public by Kepler over sixty years later; no trace of the passage could be found in the original MS of Copernicus 'rediscovered' in the nineteenth century.

Among those convinced that 'Copernicus meant not as some have fondly excused him to deliver these grounds of the Earth's mobility only as mathematical principles and not as philosophical truly averred' was Thomas Digges (p. 108). His *Perfit description of the Coelestial Orbes*, in which he set out his reasons for the unconditional acceptance of the Copernican hypothesis, was first published in 1576; but twenty years before this Robert Recorde (p. 101) makes the 'Master' in his dialogue rebuke the 'Scholar' for rejecting as 'vain phantasies the Earth being in motion at a distance of nearly four million miles from the centre of the World'. The scholar and others like him lacking sufficient knowledge of the matter 'may later peradventure be as earnest to credit it as you are now to condemn'. Nowhere more than in England was Copernicus, at least openly, given a more sympathetic consideration or to make more converts; but it is important to know what they were accepting. Digges was the only one to expatiate on the question; since by 1605 there had been at least seven editions containing this manifesto without any attempt at suppression or refutation, it may reasonably be inferred that it was fairly representative of the opinion of those competent to comment. Its *widespread* acceptance must not however be regarded as guaranteed by the number of editions, since the 'Perfit description' was only an addition to his father's popular almanach with undisguised astrological 'tokens' and 'aspects'. Digges' enthusiasm for the Copernican system was based mainly on the inadequacy of the Aristotelian cosmology, on the greatly enlarged dimensions of the world in relation to the size of the Earth derived by Copernicus by means of plausible mathematical arguments; and on the removal by him of the 'absurdities' of the equant

circles and irregular motions. Of these the second alone will commend itself to the modern reader; and though the actual scale was
still far too small Digges adopted a hypothesis—it could have been
no more—that Copernicus seemed reluctant to commit himself to:
this was the affirmation of infinite depth of the 'eighth sphere',
throughout which Digges saw the stars as more or less evenly distri-
buted. He also praised Copernicus' abolition of the 'ninth' and
'tenth' spheres—admirable indeed; but in the contemporary state of
dynamical knowledge he could not know that to achieve this Copernicus had replaced two imaginary 'spheres' by an equally imaginary
'third' *motion* of the Earth. Closely reasoned and open-minded (even
to the length of blaming the contemporary dogmatism less on
Aristotle than on his less cautious disciples) as Digges' exposition
is, it is, like that of Copernicus, based ultimately rather on philosophical preference than on 'positive' scientific evidence. Why should a
supplementary epicycle be less 'absurd' than an equant-point?
Because 'Aristotle himself, the light of our universities, hath
taught us: The motion of simple bodies ought to be simple'. Motion
in general Digges discusses ingeniously indeed; but it is still in
terms of the Aristotelian categories of straight up or down from a
centre or to a centre characteristic of the sublunary elements; and
the eternal circular motion of the celestial *about* a centre. Once the
Earth is regarded as one of the planets its (circular) motions are
'natural' as is the (circular) motions of the bodies on its surface;
so Ptolemy's rejection of a diurnal rotation of the earth whereby a
hurricane would be engendered, sweeping everything away, is
shown to be groundless. Finally, Digges returns several times with
evident relish to Copernicus' 'hymn to the Sun', which in one place
he paraphrases thus: 'For in so stately temple as this who would
desire to set his lamp in any other better or more convenient place
than this from whence uniformly it might distribute light to all,
for not unfitly it is of some called the lamp or light of the world of
others the mind, of others the ruler of the world. . . . Trismegistus
calleth it the visible god. Thus doth the Sun like a king sitting in his
throne govern his courts of inferior powers.'

Nothing in the above assessment of the Copernican revolution
is to be regarded as derogating from the imaginative genius and
intellectual courage of Copernicus and of those few—very few—
who were willing to 'stand on his shoulders'. What has been attempted
is to cast grave doubt on the acceptance of the year 1543 as 'the'
origin of the 'Scientific revolution'. It is indeed difficult to resist the
judgment of Neugebauer that in failing to make use of the celestial
latitudes of the planets Copernicus missed a fact of cardinal importance—that the planetary orbits are co-planar with the Sun; and
for purposes of visualisation his model is actually inferior to that of

Ptolemy. Also it is at least arguable that, insofar as the Copernican model had removed the 'irregular motions' inherent in the Ptolomaic and had incidentally provided the basis for somewhat more accurate planetary predictions, it may have actually *retarded* the advent of 'modern' astronomy. The reader having some knowledge of astronomy will be aware that the eccentricity of the orbits with respect to the Sun was an inevitable consequence of the fact that the Sun is not at the *centre* of the respective (true) *elliptic* orbits but at one of the *foci*. Similarly the separate angular velocities about the equant-point represent the *fact* that the angular velocities of the planets are *not* constant, but vary according to their continuously changing distances from the Sun. Copernicus might congratulate himself on his crowning achievement of the removal of the equant; but Kepler had (temporarily) to replace it before he could proceed to a higher level of precision. Until astronomers could get out of their systems the dual Platonic-Aristotelian dogma incorporated in the demand for uniform circular motion there was no hope of anything but ingenious tinkering. Copernicus was even more a slave to this dogma than had been Ptolemy. It was Kepler who, as the title of his greatest work claimed, created a *New Astronomy* (*Astronomia nova* . . .). But before he was old enough to set about this task the Copernican revolution was to be given a new twist and two fundamental postulates of Aristotelian *physics* to be shattered beyond recall.

It was highly characteristic of 'science' in the Renaissance that it was on his return from the laboratory where his passion for alchemy found its outlet that the young Danish nobleman Tycho Brahe made the observation that led to the gradual disintegration of the Aristotelian world-view: this was of a 'new' star—'supernova' as we should call it. Tycho was not the first to observe it; it was so brilliant as to attract attention all over Europe; but he alone guessed, and went on to prove, its great significance. In his first announcement (printed in the following year, 1573) he called it 'a miracle indeed, either the greatest of all that have occurred in the whole range of nature since the beginning of the world, or one certainly that is to be classed with those attested by the Holy Oracles, the staying of the Sun in its course in answer to the prayer of Joshua and the darkening of the Sun's face at the time of the Crucifixion. For all the philosophers agree, and the facts clearly prove it to be the case, that in the ethereal region of the celestial world no change in the way either of generation or corruption takes place.'

Note the order of priority of the 'authorities' appealed to in a technical mathematical work written by a man shortly to become Director of an observatory previously unparalleled in Europe, and thirty years after the Copernican 'revolution'. But Tycho's conclusion, reached by mathematical deduction from observations without the intervention of any theories, was as 'modern' as his historical criteria were antique: it was to the effect that an absolutely new celestial body was present in the 'immutable' sphere beyond Saturn. Thus was the first cosmological postulate shattered; the observations by which Tycho was driven to shatter the second were made during the year 1577, when a singularly prominent comet set Europe agog.

There was of course nothing new in the appearance of a comet. Their apparently ephemeral nature led Aristotle to the cautiously expressed view that they were formed by some kind of disturbance in the outermost sublunary sphere of 'fire'. Tycho's calculations, based on continuous observation of the comet of 1577, forced him to the conclusion that it had 'nothing in common with the elementary world, but was shown to have a motion in the Aether far up above the Moon'. This might not have been very serious; what was startling was the descent of the comet, not indeed into the 'sublunary sphere', but nearer to the Earth than the planet Venus; that is, its 'sphere' had to intersect those of the planets. These crystalline spheres—an essential feature of the Aristotelian physical cosmology—incorporated into every mathematical model since the Muslim had transmitted Greek astronomy to the West, and given conventional recognition by Copernicus in the title (*Orbium* coelestium) of his book, had in effect been ignored by most astronomers. Tycho for the first time dismissed them *explicitly*—'the circuits of the planets being wholly free and without labour and whirling round of any real spheres at all, being divinely governed under a given law'. The confused state of Renaissance thought and the character of the man himself are revealed in his decision to have the best of both worlds—the Copernican and what was left of the Aristotelian. From the former he borrowed the device of making the Sun the common centre of the planetary orbits, except that of the Moon, and, wedding it to the Aristotelian terrestrial physics of a 'sluggish element' incapable of any but 'forced and unnatural' motion, he retained the Earth (with its attendant Moon) as the centre of the World about which the Sun (with its attendant planets) revolved. Such was the model by which the greatest 'Western' astronomer since Ptolemy (or even Hipparchos) claimed to have 'avoided alike the inconsistencies of the ancient [Ptolemaic] and the physical absurdity of the modern [*sc.* Copernican]'. And there was indeed much to be said for it: it accounted perfectly well for the awkward *facts* that Venus (and Mercury, seen only with difficulty in northern countries) had never been observed at more than a small 'elongation' from the Sun, and that Mars appears to approach the Earth more closely than does the Sun. For Tycho, a somewhat virulent Lutheran, it had the additional advantage that it did not fall foul of Holy Writ—a feature that would have been of rather less consequence to the Roman Catholic canon, Copernicus, whose faith did not require the paramount authority of a literal interpretation of the Bible.

Tycho Brahe's astronomical works (except the small monograph *On the New Star*) were all *printed* at his private press on the Danish island of Hven. Copies were privately circulated, but only one—his astronomical correspondence—was literally published before his death in 1601. Thereafter, printing of appropriate preliminary matter having been completed at Prague (where he ended his days as Imperial Astronomer to Rudolph II), they were finally published at Nürnberg and Frankfurt. This delay may partially account for the comparative neglect of his system which would surely have appealed to those who, though unwilling to accept the 'physical absurdity' of a moving Earth, would have welcomed a more accurately based system than the work of Copernicus afforded. Be that as it may, what is of far more importance in the present context was the foundation of the observatory at Uraniborg. Despite an occasional hint to the pope or some other great prince insinuated in a dedicatory epistle, almost exactly a century had passed before the well-conceived plan of Regiomontanus (p. 38) had once more been put into effect, and on a scale never previously approached in Europe. It is an indication of the lack of communication between the West and the Middle East that the remarkable achievements of the Moslems consequent on elaborately equipped and competently staffed observatories at Baghdad and elsewhere were unknown.

Though almost universally obsessed by the supposed significance of astrological portents, those princes who had the means to establish a permanent observatory failed to realise the inconclusiveness of predictions based ultimately on observations made without any serious concern for maximum accuracy. William IV, Landgrave of Hesse, seems to have set no reliance on judicial astrology (p. 34); yet the science of the heavens had for him so great an appeal that about 1560 he made extensions to his castle at Kessel, equipped the building with the best available instruments, and appointed a distinguished astronomer to carry out systematic observations; perhaps most significant of all he frequently took part in the work himself. When the young nobleman, Tycho Brahe, fired by the Landgrave's example, expressed the intention of devoting his life to astronomy his seniors were scandalised; the intervention of the Landgrave with Frederick II of Denmark on behalf of the promising young astronomer was probably decisive in the promotion of an enterprise on a far greater scale than that of his own pioneering achievement.

With surprising magnanimity the king placed the island of Hven at the absolute disposal of Brahe, and to the modest fortune derived

from the latter's estates added a subvention calculated to provide a
building of palatial dimensions, a large staff of skilled assistants and
numerous instruments of the most elaborate design. The well-known
and somewhat idealised picture of the great mural quadrant included
in one of Brahe's books is one of the most important iconographical
documents of Renaissance science; the vivid glimpses of the activities
of the observatory depicted therein reveal two of the necessary
conditions of modern science: the continuing striving for greater
quantitative precision and a degree of 'socialisation' to provide the
means. Since the former involves the emergence of an almost independent science of metrology its further consideration will be deferred
to the next chapter.

   The emotional drive behind this immense and sustained endeavour
of Tycho Brahe was the hope of creating a new science of astronomy.
It is one of the ironies of the history of science, by no means unique,
that when the unprecedented series of observations of the planet
Mars *all round the orbit* were bequeathed by him to his loyal associate
Kepler, in the expressed hope that by their use Tycho's system might
be fully confirmed, the 'new astronomy' that actually emerged
relegated not only the Ptolemaic and Copernican, but also the
Tychonic, systems to the museum of antiquities. How far this upshot
is to be considered a part of the 'Copernican revolution' is a matter
of taste; but for about fifty years both before and after the publication of Kepler's *Astronomia nova* (1609) the question exercising the
minds—and also the souls—of European intelligentsia was less the
nature of the solution of the celestial problem than what it amounted
to when obtained.

   For one prophetic soul at least there was indeed no doubt: in
a series of dialogues written in Italian and Latin Giordano Bruno
asserted unequivocally that the planets we see and the Earth on which
we live circle round the Sun; that the Sun is merely one of an infinite
number of similar stars whose planetary companions are invisible
because of their immense distance from us; that all these motions
take place in an infinite space that has neither circumference nor centre
(alternatively that every point in it is both circumference and centre);
and that the attributes of the World are uniform throughout—the
distinction between the 'celestial' and the sublunary elemental realms
being vehemently rejected. He asserted a great deal more about the
nature of the World; but in doing so he drew on the hermetic and
magical tradition that is the subject of our Chapter 11.

   From the day in 1600 when he perished at the stake on the Roman
campagna Bruno has been regarded by some as a 'martyr of science',
by others as a disreputable babbler whose influence on science was
negligible and who deserved all he got. That he was a 'martyr' of
science can be summarily disposed of; his knowledge of science—

even of that of his own day—was very limited. He was suspicious, even contemptuous, of mathematics as a means of gaining *natural* knowledge; and his bold assertions were based on nothing more than inspired guesses prompted by an extensive if not very scholarly knowledge of ancient and medieval natural philosophy and the broad lines of the Copernican and Tychonic innovations. Unorthodox as his 'science' undoubtedly was, it was for his fundamentally theological standpoint that he was destroyed; in the temper of the times it is difficult to see what else his superiors could have done, though the treachery and unnecessary cruelty of the circumstances make it one of the less edifying episodes in the history of the Roman Church.

Of his actual or possible influence on the thought of his contemporaries more will be said later; but with the other judgment of history—that he was a disreputable babbler—it is however necessary to deal at this stage. Disreputable he may have been—there were many such figures in the sixteenth century whose profound influence modern historical research has been at pains to reassess—but a mere babbler he certainly was not. Kepler, at least, didn't think so, even though he rejected most of Bruno's assertions by arguments soundly based on the limited knowledge of the age. In a letter to Galileo he expressed shock at the absence of any reference to Bruno in the former's epoch-making *Starry Message* (*Nuncius sidereus*) in which had been demonstrated the immense enlargement of the visible universe by the discovery of countless 'new' stars and of the subordinate 'planetary systems' represented by the 'Medicean stars'—the satellites of Jupiter. Moreover Bruno's rejection of mathematics as an essential aid to the revelation of the nature of things was far from being a merely frivolous prejudice. When Galileo claimed 'the book of Nature is written in the language of mathematics' he showed by the examples he adduced that by 'mathematics' he meant 'geometry'. But Bruno correctly insisted that theorems of geometry express relations between *ideal* entities which, being figments created by the human mind, could reveal nothing about the *real* world. Had his judgment been generally accepted there would of course have been no 'scientific Revolution'; but equally the world might have been spared the persistent output of sophisticated nonsense generated by the application of impeccable mathematical techniques to confusedly conceived or even false 'data'.

As in the 'revolution' ascribed to Copernicus so in the cosmology of Bruno there was nothing really new. Inspired above all by the liberation of the human spirit effected by the cosmical relativity (chastened by 'learned ignorance') of Nicholas of Cues Bruno based his cosmology on a reappraisal of the atomic theory of Demokritos and of the 'cosmical harmony' of Plato. By contrast he drew from Aristotle a respect for the 'solid ground of nature' and a recognition of the perils inherent in the dominance of mathematics. Bruno was thus an eclectic—generous in praise, unsparing in criticism, of those pioneering spirits on whose work he had drawn —but the selectivity of his eclecticism yields a creative synthesis in which the earlier views are given a new significance. In initiating the astronomical 'revolution' Copernicus performed the essential task of showing it to be academically respectable; Bruno's no less essential function was to put the cat among the academic pigeons. And behind the antics of this provocative animal it may be possible to discern the 'prophetic soul of the wide world dreaming of things to come'.

E

I O

ACCENT ON QUANTITY

A recent tendency to see the 'end' of the Renaissance and birth of the Modern World in a greatly heightened concern with the quantification of the data of experience undoubtedly represents an important insight, but it can be misleading. There was nothing fundamentally novel in the activity of measuring and tabulating; the Akkadian–Babylonian civilisation of the third millennium B.C. achieved a level of recording and calculating whose sophistication relative to the general culture has probably never been surpassed. But it was the monopoly of a bureaucratic and ecclesiastical caste having the laudable aim of ensuring a reserve of food against a rainy day—or more likely an unrainy year—with the consequent necessity of being able to pin guilt on any primary producers who tried to cheat the temple 'god'. Nearly two millennia later from roughly the same region came the ingenious zig-zag functions for forecasting the position of the planet Venus in a manner somewhat suggestive of the kind of calculus used today in quantum mechanics. But in all this there was very little hint of what Plato admirably called 'wonder'—wonder in respect of natural knowledge—or of inventive ingenuity in the search for new techniques of 'quantity-surveying'.

Greek 'quantification' took a totally different course, being concerned only incidentally with measurement or counting, and with number only as a relation or 'ratio'; which reveals how inappropriate was the choice of the word 'numeracy' as a measure of mathematical understanding. The famous inscription over the portal to Plato's Academy—'let no one enter here who cannot geometrise'—represented a cultural ideal, not, as it might to us, an intellectual exclusiveness. Mathematics rather than 'numeracy'—the Greek numerical notation was even more inferior to the Babylonian sexagesimal base than is our own—was an essential part of the Greek way of life. This, as we have seen (p. 52), was fully realised by the creative genius of the *quattrocento*.

The Greeks had done their work so supremely well that there was not much the Renaissance mathematicians could do to extend it. The achievement of the *cinquecento*—algebra and the algebraicisation of relations in the plane and spherical triangle which at the end of the century came to be known as 'trigonometry'—grew not so much out of 'Greek' mathematics as out of the special interests of Hellenistic mathematicians—Archimedes, Hipparchos, Heron, Ptolemy, and those last 'flashes' of ancient genius, Pappos and Diophantos.

It is unnecessary to accept unreservedly the Marxian conception of history to recognise that the need for quantification was thrust upon Renaissance society not so much by the intellectual curiosity of academics as by the logic of events—political, military, industrial and commercial. In no contemporary testimony is this more tersely put than by Nicolò Tartaglia in his book *Various questions and inventions* published in 1546 and dedicated to Henry VIII. The versified preface *To the Reader* opens as follows: 'Whoever desires to behold new inventions let him not take (*tolte*) them either from Plato or Plotinus or from any other ancient Greek or Roman, but only from craftsmanship (*larte*), measurement and theory (*Ragioni*).' Tartaglia might not unfairly be described as a 'failed academic'. As a consequence of early poverty he lacked literary education; and as a consequence of the mutilation of his face by French troops at the sack of his birthplace, Brescia, when he was eleven years old, he suffered from an impediment (*tartaglia* = stammerer). The highest rewards of scholarship were thus denied him; his translations of Euclid and Archimedes were less than adequate, but the necessity to apply his mathematical virtuosity (p. 132) to military problems guaranteed him an honoured place in the history of science. In seeking the patronage of Henry VIII he artfully put the first 'question' into the mouth of Francesco Maria della Rovere, Duke of Urbino, with an indication that he was already on good terms with a prominent ruler.

The special circumstances of Tartaglia's pronouncement impose caution in drawing any general conclusion: we can but note that his most important advance in the *theory* of projectile motion—a subject debated with rational insight and vigour in the Faculty of Arts in Paris in the thirteenth and fourteenth centuries—appeared in the course of a discussion on the improvement of the accuracy of artillery. This was the clear statement illustrated by a diagram that, contrary to the generally unquestioned assumption of Aristotle, the projectile does *not* move at first along the line of projection but begins its natural fall towards the Earth immediately it leaves the barrel. Unfortunately Tartaglia provides no systematic measurements to verify this assertion that he arrives at deductively; the fact, if it is a fact, being clearly not directly visible. We cannot even be absolutely sure that he took part in any gunnery trials such as those reported by the Duke of Urbino to have been made with a culverin at Verona. The discovery itself was, however, an indispensable basis of Galileo's mathematical demonstration of the parabolic trajectory of an unresisted projectile.

It does not appear that the science of gunnery was at all influenced by Tartaglia's perfectly correct 'invention'; master gunners were far too steeped in the traditional lore to be influenced by a 'theoretician's' ravings that were obviously at odds with common sense. And for their purpose they may well have been wise; the assumption of a uniformly horizontal path of a ball fired at point-blank range was not only 'common sense', but the Duke's culverin, when aimed horizontally at 200 paces, did in fact hit a mark on the same level. Tartaglia was right in objecting that for such a short trajectory the vertical deviation from a straight line was beyond the power of human vision. The large sources of error consequent on the absence of rifling and lack of uniform bore of the barrel would in any case have introduced confusion into the interpretation of systematic trials. The lesson to be learnt from this 'case-history', admirably set forth by Alexandre Koyré, is two-fold: that whereas the application of immature theory may be positively harmful to a complex craft based on traditional experience, conversely progress and refinement of theory may be impeded by too strict a reliance on empirical verification. Tartaglia's achievement was to have produced the first printed treatise on ballistics (*Nova scienza*) and, in the later *Quesiti et inventioni diverse*, corrected a universally accepted but erroneous principle. His limitations were revealed by the artificial expedients he resorted to in the effort to harmonise his correct principle with the discrepancies of crude experience. The subsequent achievement of this latter task, as Koyré reminds us, was an insufficiently recognised mark of the genius of Galileo. How far the latter's indispensable contribution to the 'scientific revolution' was related to technological and social conditions it is impossible to say. Though at the beginning of the *Discourses and mathematical demonstrations of two new sciences*, in which the decisive part of this contribution was established, he refers in a familiar way to the work going on in the great arsenal at Venice, and in the earlier 'discourses' he is concerned with correcting the traditional assumptions about loaded beams and problems of scale, the two discourses on the question of 'local motion' (the term also used by Tartaglia) contain hardly any reference to gunnery but concentrate on the fundamental problem of the accelerated motion of freely falling bodies, 'be the cause of the motion what it may'. It is this abstraction of the formal relation from the particular instances that makes possible the complete 'quantification'; this together with the introduction of the 'thought-experiment' constitutes the essence of the 'mathematical philosophy'. Neither of these was wholly novel: the schoolmen of Paris and Oxford had closely approached them; Galileo, but not only he, took the decisive step of seeking confirmation of the theory, not by *crude* experience, but by specially designed experiments in which all but the formal relationship was as far as possible excluded.

However little influence the two works of Tartaglia may have had on the master-gunners, they aroused enough intelligent interest in the problem to warrant an English translation of the first three 'books' of the *Quesiti*. Published in the Armada year (1588) under the title *Three bookes of colloquies concerning the arte of shooting in great and small-peeces of artillerie* . . . the translation was carried out by Cyprian Lucar who added a long appendix of citations from other authors and twelve years later published a largely original work, *A Treatise named Lucar Solace*. Lucar, a scholar of Westminster School and of New College, Oxford, was a bencher of Lincoln's Inn in 1568; these facts, taken in conjunction with wide technical reading displayed in the appendix, provide evidence that it was not from the 'practical men' nor even only from the professionally concerned 'arts men', such as Thomas Digges in his revision (1579) of his father's *Stratiotikos*, that a new approach to the problems of ballistics was in fashion before the actual battle revealed the marked superiority of English gunnery to the Spanish.

Mention of the long artillery duel that preceded the final destruction of the 'Invincible Armada' should remind us of Garrett Mattingly's opinion that technological progress in gunnery had brought to birth a new era in naval warfare in which 'nobody in either fleet knew how to fight a "modern" battle. Nobody in the world knew'. Before 31 July 1588 naval battles were fought almost entirely on the decks of ships grappled together to permit of boarding; guns were fired mainly against men, not ships. All this was changed. The point-blank range of 200 paces of which the Duke of Urbino's culverin had been capable, was, so it was being said, raised to 700. Owing to the variety of 'paces' and gun bores this statement must not be taken literally but only as an indication of the change of scale; the wary manner in which the rival commanders at first faced one another's broadsides is perhaps the best evidence. Should a few of the thirty-pound shots that the culverins could hurl strike below the waterline they might, so it was thought, render a great ship unmanageable, or even sink her. The events of the first day's fighting showed how unreliable were the estimates of fire-power:

the range had to be much closer, when accuracy of aim became subordinate to speed of fire. Nevertheless, though 'science', despite all the textbook figures of quadrant readings to ensure accuracy of aim, had achieved little in this respect, the *belief* in greatly enlarged fire-power was sufficient to revolutionise tactics in one of the decisive battles of world history.

Had English gunnery been the sole factor involved the Armada would have been unpleasantly shaken but far from crippled. But another factor relevant to our topic of quantification might have brought about a decision even without the fireships and northern storms: both fleets were running out of ammunition. The English were able to replenish theirs after a serious delay; but the permanent lack of heavy projectiles in the Spanish ships made it possible for the English to approach them at short range, inflicting heavy casualties and rendering the ships far less seaworthy to face the storms of the subsequent homeward passage.

The problem of military logistics was not new in 1588: the French had had to estimate the needs of extensive artillery trains operating beyond the Alps as early as their 'Italian venture' of 1494. But in 1588 no one had any idea how much to allow for the loss due to moving targets at relatively long range; perhaps also they had forgotten that in a sea fight the missed shot could not be retrieved and used again! The influence of these logistic problems on the production of mathematical textbooks is all too plain: the obsession with the extraction of square roots was clearly bound up with the calculation of the number of pikemen and the extent of the rank in the infantry formation, e.g. when three 'squares' had to be formed out of a given number of men. More sophisticated was the mathematics of fortification, especially from the time of Sammichele, whose re-entrant walls affording cross-fire can still be seen at Verona and whose theory was first published in a textbook by Dürer.

In no case is the technological factor in the progress towards the quantitative basis of modern science more strikingly illustrated than in the life of Simon Stevin of Bruges. Lacking any youthful academical training but probably aided by early concern with 'numeracy' as a merchant's clerk he was the first to bring about a wide recognition of the importance of replacing the clumsy 'vulgar' by decimal fractions. It is a regrettable anatomical fact that the first 'digital computer'—the human hand—has *five* digits. Had there been six a reasonable compromise might have been reached with the far more rational 'Babylonian' sexagesimal system, based on the first three primes, than by omitting '3' as in the greatly inferior decimal system. Stevin's contribution was the publication of a systematic textbook; his clumsy notation for place-value was soon replaced by some kind of 'point'. His book—*De Thiende*—was undated but published probably in 1585; yet, though the first English version of the Flemish original appeared in the following year, Newton still displayed his optical calculations in appalling vulgar fractions a century later. After nearly three centuries opposition among the British has not been wholly overcome.

Though his studies in arithmetic, algebra (in certain respects more 'modern' than the much more profound work of Viète, p. 109) and tables of compound interest appeared in French in 1585 Stevin's nationalistic bias in favour of his own language prevented the wide recognition of his much more sophisticated works on mechanics before the French translation by Girard nearly fifty years after their Flemish originals appeared in 1586. In these works, inspired by the problems he had to face as engineer and quartermaster general in the service of Maurice of Nassau, Prince of Orange, he gave the first experimental and analytical demonstration of the paradoxical pressure of water columns (Pascal's Law) on which the hydraulic press was later based; of the triangle of forces in its statical form; and of the equilibrium of bodies on inclined planes. Both statical discoveries provided new techniques of calculation; but the law of the inclined plane is of special interest: it is an early example of a 'thought-experiment', and implicitly involves the concept of virtual work. All this, and a good deal more in a wider technological setting but supported by quantitative demonstrations, was published by Stevin when Galileo was only twenty-two. It is no derogation from the genius of the latter (his mind was already engaged on similar problems) to emphasise that in those very methods for whose deployment Galileo is often held to be the 'founder' of modern science—analysis in quantitative terms, 'thought-experiments', and actual experiments with specially fabricated apparatus—he was anticipated by a man apparently largely self-taught. Stevin had certainly had none of the advantages (nor admittedly the disadvantages!) of a university education before he had established a talent for invention. A possibly essential factor in Galileo's achievement was his early application of academically acquired mathematics to practical problems of surveying and drainage, etc. In Stevin's case the factor is, as it were, stood on its head: the technological demands posed by late Renaissance society elicited the same methodological emphasis critical for the 'scientific revolution'. Had his works been published simultaneously in Latin or even, as in Galileo's case, in one of the more widely known vernaculars, 'popular' history of science might have placed the emphasis on the two men rather differently. If it be urged that the 'popular' image of Galileo is of the 'astronomer' the point is well taken; but in the revelations of the telescope broadcast in the *Starry Message* there is hardly any mathematics and not very much 'quantification', the sensational message being of a new dimension of *quantity* in the cosmos. The truth seems to be that by the 1580s 'modern' science was 'in the air'; to look for a single 'founder' is to risk seeing a mirage.

The exigencies of military operations involving long lines of communication may also have called forth a striking advance in the precision of topographical maps. The marine 'portolan' charts based on astronomical coordinates were even before the sixteenth century articles of commerce (p. 31); but corresponding representations of landward areas larger than those of monastic estates were extremely rare, and for the most part merely decorative. In 1585 appeared the first of the three parts in which Mercator's world *Atlas* (his term) was subsequently completed ten years later, comprising altogether 111 maps. The huge increase of reasonably accurate topographical knowledge had been made possible by the invention by his teacher, Reinerus Gemma, of the process of triangulation, whereby, from a single accurately measured base-line, the position of any feature could be calculated. The method—the basis of all subsequent cartography—appeared in 1533 as an 'addition' to Gemma's edition of the famous *Cosmographia* of Peter Apian. The apparently rather slow application of triangulation seems to have been due to the absence of a suitable instrument combining a levelling device, firm stand, and with sights movable in both altitude and azimuth. This instrument was first referred to as a 'theodelite' [*sic*] by Thomas Digges in 1571. The improvement in cartography by which Mercator is best known, his cylindrical projection, was of less significance at the time, since the distances involved in land survey were usually too small to be seriously affected by the Earth's curvature, and it was too difficult for seamen to apply until more adequately explained in textbooks: the first of these was written by Stevin in 1599 and translated into English by Edward Wright as the *Haven-Finding Art*.

No aspect of the quantification in the Renaissance is of more general interest than the technical aspects of the well-known rise in the price-level during the sixteenth century. The cause of this—even the facts themselves—is still a controversial topic among economic historians; we are concerned only with one very significant but insufficiently known contemporary attempt to analyse it in quantitative terms, and on the other hand with the technological response.

The analysis was in fact prompted rather by the evils consequent on a local debasement of the currency than on the problem of a general change in the price-level as such—a state of affairs hardly recognised as early as 1523 when Copernicus made his report at the request of Sigismund I of Poland. The principality of Ermland had during 1519–21 suffered serious devastation as a consequence of one of the periodic struggles between the Polish natives and Teutonic Knights on their eastern border. In the following year

Copernicus made a preliminary survey, on the basis of which he was able to demonstrate that the debasement of the currency only increased the evil of rising prices. From the historical point of view the report is of much more interest than would be that of a merely *ad hoc* 'crash programme'. He shows himself to be aware of the whole basis of currency as a medium of exchange; the necessity for hardening silver with copper; the distinction between the value of the metal and the money of account created from it in which, for instance, the cost of minting must be allowed for. Observing the bullion dealers and jewellers (an important guild at that time) picking out the 'good' coins for melting down he discovered 'Gresham's Law' of displacement fifty years before its namesake; recognising the dishonest sources of profit made available by variations in local standards of value he proposed a gold-based international standard. Altogether a rather impressive document. Even more remarkable is the fact that a good deal of the work had been done even earlier by the fourteenth-century genius, Nicole Oresmes, Bishop of Lisieux, who also had seriously considered the possibility of simplifying astronomical theory by giving the Earth a daily rotation. The limited 'modernity' of the Renaissance is suggested by the fact that neither of these works was printed until recent times. After 1560 Elizabeth was forced to undertake the task of undoing the evil her father's and brother's successive debasements had brought about.

The growing importance of spot-cash for the maintenance of the large armies—often embodying a high proportion of mercenaries, having none but the cash-nexus to keep them in the field—involved a marked growth in the minting of coin, hence of the extraction and assay of precious metals. The main lines of the latter had become traditional during the *quattrocento* but one of the most important bases of the subsequent *science* of chemistry was the refinement of the means of separating metals, and the availability of an instrument able to provide comparable quantitative data on the quality of ores. The assay balance and weights pictured in Agricola's *De Re Metallica* (1556) differ in no essential feature from the chemical balances and fractional weights in use in all but the most accurate chemical analysis until the fairly recent past. A remarkable testimony to the wider significance of quantification involved in assay is the startled perplexity of the Assay Master, William Humfrey, on learning that a hundredweight of iron had been 'converted' into a much greater weight of copper. Already assuming the much later established principle of the conservation of mass in chemical change, but ignorant, as was everyone for a further two centuries, of the more sophisticated law of chemical equivalence, he could offer no explanation. Perhaps most important of all, as Professor M. B. Donald (*Eliza-*

bethan Monopolies, 1961) suggests, 'He had found by experiment something far more important than what he was looking for'. The English metallurgists were unaccountably ignorant of the appearance in 1574 of Lazarus Ercker's book on mining and minerals, revealing a considerable advance on the methods of refinement described by Agricola.

Metal assay was not the only kind of chemical activity in which recourse to quantitative methods became more marked; by the introduction of *graduated* glass cylinders 'uroscopy' was aiming at becoming 'urometry', though what physicians thought was to be learnt by this greater refinement remains problematical. Some insight (at least as old as Galen) warned them that the characteristics of urine were not those of a *mere* waste product, but in some way indicative of the disturbed 'humoral balance' of the patient: the result was an extensive literature, dating back to the Roman physicians, of signs and portents, admirable indeed as a training in accurate observation but matched by an involved diagnostic scheme that was almost wholly imaginary. Much more promising was the refinement following the revived interest in the Roman cult of mineral springs. Towards the end of the sixteenth century this refinement was taking the form of quantitative assay of the *solutes* present in the waters, accompanied by some attempt to classify them by qualitative tests. A good case has been made out for regarding this enterprise as playing a no less important part than metal assay as the basis of chemistry. That Robert Boyle profited by knowledge of this primitive form of analysis there is now little doubt.

Thus far an attempt has been made to provide evidence for a widespread and growing concern with quantifying the data of experience. Of more importance—at least until the establishment of the 'new experimental philosophy'—was the discovery and application of the rationale of measurement itself; not surprisingly this took place in the practice of astronomy whose very existence had always depended on the greatest possible attainable accuracy, and whose progress had been virtually halted for want of greater precision. In achieving this Tycho Brahe, if not alone, was certainly pre-eminent. With previously unparalleled material resources at his disposal he was able to install and permanently maintain instruments of the finest workmanship, hire sufficient skilled assistants to maintain continuous observation, and subdivide each task to attain the maximum precision in timing and recording. And beyond all this he recognised the *inevitable* presence of error of different kinds and the consequent necessity for calibration of the instruments that is, the removal of all the known sources of error (e.g. instability of the foundations, bending of metal rods under their own weight) and thereafter the performance of such routine checks as enabled him to estimate

residual error and apply suitable corrections. This process of calibration applied especially to the measurement of time. Clocks were
by then a commonplace; but no method was known of maintaining
a 'constant' rate for more than a few minutes; to minimise error the
clock would be calibrated by observation of some standard celestial
object shortly before the event to be studied was expected to occur.
By the inclusion of a diagonal scale the precision of the graduation
marks on the great mural quadrant (p. 127) was increased to such a
degree that in the absence of telescopes the ultimate limit of precision
was set not by the instruments but by the limited resolving power
of the human eye—of the order of a minute of arc at best. It was this
order of accuracy that enabled Kepler to make his dramatic claim:
'These eight minutes showed the way to the renovation of the whole
of astronomy.' 'These eight minutes' (of arc) constituted the discrepancy between the observations made at Uraniborg and the best
theoretical 'models' constructed on the 'old' assumptions—Copernican or Tychonic. But the new standards of measurement made
impossible a discrepancy of such magnitude. So the 'old' assumption
of uniform circular motion was at last proved to be untenable.

Implicit in Kepler's claim there is however another assumption
—the relation of a set of quantities to their average. That a more
sophisticated attitude towards this assessment was becoming prevalent even in the transactions of political economy is well brought
out by Professor J. U. Nef in his important work, *The Cultural
Foundations of Industrial Civilisation.* When the clerks of the excise
were asked to provide a statement of the export of coal for seven
successive years they submitted a *total* figure for the seven-year period.
When the statement was returned for correction they broke down
the figures year by year; nevertheless one of them, as if in protest,
noted that '13651 is the 7 part of the coles transported [overseas]'!

A comprehensive study of the origins of statistical science has
probably still to be made; it would be no easy task, since it embodies
a manifold of factors: collection, that normally involves *selection*,
of the data; hence a conscious critique of sampling and what infer-
ences can be drawn from such incomplete data; and this in turn
involves the conscious application of the special branch of pure
mathematics known as the theory of probability. Statisticians seem
generally to regard as the pioneer work of their craft the *Natural and
Political Observations . . . made upon the Bills of Mortality* published
by John Graunt in 1662. Also it has been common to assume that
the abstract theory of probability had its origin in a correspondence
between the mathematicians, Blaise Pascal and Pierre Fermat, on the
problem of the division of the stakes in an uncompleted game of
chance. Professor Oystein Ore has however provided conclusive
evidence that the 'typical Renaissance Man', Girolamo Cardano

(p. 52), had written, but not published, a small treatise on the esti-
mation of the chances in various games about a century before the
generally accepted—and of course more profound—discussion
between Pascal and Fermat. The importance of Cardano's treatise lies
in the fact that he sought *general* rules applicable to the common
factors (e.g. serial repetition of a given throw with dice, or a draw
from a pack of cards) in superficially very different games. He
discovered the 'power law' applicable to repetition, which Ore
thinks ought to be called 'Cardano's Law'. His clear recognition that
the chances of throwing a given point with six 'blind' dice (one face
only of each bearing a number) are not identical with those in
respect of a single die is equivalent to an appreciation of the so-called
'law of large numbers', where averaging is involved. There is no
evidence that the work had any direct influence on his contemporaries,
but on his own admission Cardano was an inveterate gambler and
his opponents included men in high office; it seems hardly likely
therefore that no discussion took place. His sagacity is shown by his
likening the habit to an incurable disease—a proper subject of
discussion between medical doctors; it is certain that for his own
case he found not even an alleviation of the symptoms!

The year dividing the centuries (1600) saw the publication of yet
another book dividing the science of the Renaissance from modern
science. Though he made little if any use of mathematics, William
Gilbert, physician to Queen Elizabeth, based his most important
demonstration on a set of simple *measurements* of the successive
positions of a mounted compass needle when it was moved over
the surface of a spherical lodestone (natural magnetite) representing
the Earth. This was an extension of the observation, made some
years previously by the instrument maker Robert Norman, that a
compass needle magnetised *after* it had been balanced 'dipped'
towards the north at a constant angle. Further reports had indicated
that the 'dip', like the 'variation' (now called 'declination' by
physicists), varied from point to point on the Earth's surface; but
whereas the change in variation was rather irregular, the magnitude
of the dip was roughly proportional to the latitude, becoming less
as the equator was approached. Gilbert saw in this behaviour a
clue to the magnetic nature of the Earth itself. The nature of the
demonstration is expressed in the title of his work: *On the Magnet,
magnetic bodies and the Great Magnet, the Earth, a new physiology,
proved by numerous arguments and experiments*. The other aspects
of this epoch-making book will be referred to in Chapter 12.

FROM MAGIC TO SCIENCE

'And thus much concerning magic; whereof I have both vindicated the name itself from discredit and separated the true kind from the false and ignoble.' Francis Bacon's 'vindication' of the name of magic consisted in reminding readers of his *De Augmentis Scientiarum* (an enlarged Latin version of the earlier *Advancement of Learning*) that the three 'wise men' who came from Persia to worship the new-born Jesus of Nazareth had been known ever since as 'Magi' 'for among the Persians magic was taken for a sublime wisdom'. But in the recent past, before his own day, Bacon went on to say, the term had fallen into disrepute as the practices to which it referred had degenerated in a number of ways, which he proceeded to describe.

At various stages in this book it has been necessary to emphasise both the paramount influence that judicial astrology (p. 34) had on the advancement of astronomy in the Renaissance, and also the reasoned protests generated by the artificial assumptions underlying the more notorious and personal forms of astrological prediction. On any account, astrology and magic shared a common fate, in becoming storm-centres of controversy in the sixteenth century. In this chapter it is proposed to distinguish—so far as this may be valid—the 'true' kind of magic from the 'false and ignoble' and to examine the role of each in Renaissance society. The task is in many ways far more complex in respect of magic than has so far been shown to be the case in astrology; but one of the results of our enquiry will be to reveal a degree of over-simplification in the useful but superficial distinction between the apparently wholly arbitrary and bogus 'horoscopic genitures' on the one hand, and the factual correlation between the phenomena of the tides and seasons and the regular sequence of motions of the heavenly bodies on the other. One of the most challenging aspects of Renaissance magic lay in the appeal to the parallel of macrocosm and microcosm—Man and the Celestial World—one consequence of which was the overlap of magical and astrological fields; if the faithful were ready to accept that 'one star differeth from another in glory' it might be—and was—correspondingly assumed that it differed also in the evil that it might inflict.

The title of this chapter—taken from Paolo Rossi's magistral work *Francesco Bacone—della Magia alla Scienza* (1957)—must not be held to imply that modern science was simply a transformation of magic, or even that magic was a *necessary* phase in the emergence of science. What has already been several times suggested and will now be developed more at large is the historical fact that many—perhaps most—of those who effected the revolutionary changes in natural knowledge—Paracelsus, Cardano, Copernicus, Bruno, Kepler, to name the most prominent—were motivated by a deep conviction of the working of occult forces beyond the range of immediate sensation. That this was an addiction to 'magic' would almost certainly have been hotly denied by Copernicus and somewhat less convincingly by Kepler. But one of the principal aims of this enquiry is the demonstration of how difficult, if not actually impossible, it is to draw any profitable boundary between the 'natural' and the 'magical'. Nowhere is this more clearly revealed than in the writings of Bacon, who in one place stated that he would 'revive and reintegrate the misapplied and abused name of Natural Magic for at least a part of Natural Prudence'; in another he speaks contemptuously of 'that natural magic that flutters about so many books'. There is here no fundamental inconsistency: he pin-points such 'credulous and superstitious traditions and observations' as 'sympathy and antipathy' whose use we now are able to see, like Molière's *faculté dormitive*, commonly did no more than restate a problem in different words. But even here we meet with an obstacle to the drawing of clear boundaries; for did not Fracastoro (p. 95) say that without a consideration of Sympathy and Antipathy it would be impossible to investigate or to demonstrate the nature of contagion?

If at one end of the scale 'natural magic' generates the 'centaurs and chimaeras' that, as Bacon reminds us, were Ixion's progeny when he embraced a cloud, at the other end it is barely distinguishable from the 'natural miracles' that, as Alberti (p. 17) and Dürer envisaged, 'men can do if they will'. That the latter was the most settled of Bacon's interpretations of the term will be demonstrated in our concluding chapter; here it will be sufficient to refer to a brief statement made in the context from which most of the above remarks have been drawn: 'I however understand it [*sc.* magic] in general as the science which applies the knowledge of hidden forms to the production of wonderful operations.' Magic is thus to be regarded as essentially concerned with the creation of 'works' rather than of mere knowledge as such.

Bacon is of course no longer regarded as the 'Founder of Modern Science'; magic—however interpreted—cannot therefore be seriously considered as a *sufficient* condition of the spectacular scientific achievement of the seventeenth century. Nor in a strictly logical sense could it be shown to have been a *necessary* condition. But history is not primarily an affair of logic: all we can do is to take account of 'what actually happened' and, to a more limited but vital extent, of what *didn't* happen. In conditions of unsurpassed mastery of mathematical, logical, and aesthetic form in classical Greece 'modern science' did *not* emerge; in the High Middle Ages—perhaps more justifiably called an 'Age of Reason' than is the eighteenth century—a new and somewhat different impulse occurred but petered out; only at the culmination of the 'Renaissance' was 'science' in very much the modern acceptation of the term brought to birth. What was the differentiating factor, the necessary condition previously absent? It is unlikely that any such neat logical isolation could be achieved; nevertheless one factor is present throughout our loosely defined period (1450–1600), and not only present but pervasively and increasingly so, that in the other two progressive ages remained very much in the background. This obsession with the possibility of allying the human mind with 'supernatural' powers to bring about miraculous changes bears a more than superficial resemblance to our present obsession with sex: both spring from the unconscious, are deeply involved in social and religious taboo, and have almost unlimited potentiality for beauty or disaster. It need hardly be added that, springing from the unconscious, they might be expected to find conscious expression in similar or even identical symbols; and, in the case of alchemy such was found to be the case, notably by C. G. Jung.

Despite this positive evidence it is not so long ago that any attempt to demonstrate the high significance of the concern with the 'supernatural' to the contemporaneous emergence of the 'scientific method' would have been regarded as frivolous if not actually 'obscene'; and not wholly without reason, since so long as mere sensory experiences correlated by the appropriate mathematical instruments were regarded as both necessary and sufficient conditions for 'real' science, the appearance of 'irrational nonsense' in the otherwise exemplary achievements of men of genius was accepted as an unfortunate but otherwise harmless aberration. When possible—and there have been and probably are still many cases where it *was* possible—these aberrations were hushed up. Historians of science—and perhaps not only of science—saw what they wanted to see, as Bacon brilliantly displayed (p. 160). Not only did they brush under the carpet what they didn't like the look of, but in the case of what William Whewell called the 'Stationary Period' of the Middle Ages they failed to see that there was anything worth looking at.

To develop this question at greater length would be out of place in this context. The age of the Renaissance, however, resembles our own in being one in which the changes in the material conditions of life, great as they were and are, will in the perspective of history appear of less significance than the 'transvaluation of all values' that accompanied them. This process, whether as cause or effect of the 'new philosophy', necessarily concerned the whole of Western society. In seeking to evaluate its causal aspect in relation to the emergence of science we shall not confine attention to those figures who would normally be regarded as 'scientists'. The wind of change bloweth where it listeth; sometimes from the most unexpected quarters it stirs the embers of those least conscious of it.

How then shall we set about the task? To encompass it volumes (six, in Lynn Thorndike's *History of Magic and Experimental Science*) have been written, but so far with but partial success. In this brief sketch an attempt will be made to combine cohesion and lucidity by taking one or two outstanding figures as representative, or perhaps more accurately as focal points, in relation to their own society and also to the relevant stresses and strains of the subsequent development. Of these figures none stands out more strongly than Iohannes Kepler. At various stages in our narrative (p. 128) we have given reasons for regarding him as the supreme example of a spirit stirred to prophecy by an imagination sensitive to all the currents of past wisdom, whether of staid fact or elaborate fancy. But so great was his intellectual integrity that he was able to transmute the shadowed dreamland of the past into the 'solid ground of nature', to which consistent appeal would be made in the age about to open. Pious to a fault he took to heart the dissensions surrounding him in the theological seminaries where he was to be prepared for the reformed ministry; yet he lived to be the object of persecution by both parties in the drama.

Before reaching the age of twenty-one, in disputations among the theological students at Tübingen, Kepler was defending the Copernican hypothesis to which his distinguished teacher, Michael Maestlin, often referred. There was, of course, no objection to this: the office of *Advocatus diaboli* was a necessary aid to apologetics. Six years later (1596) when the young man proposed to publish a book embodying wide-ranging and somewhat startling speculations on the *Mystery of the World* (*Mysterium cosmographicum*) he received a friendly warning from the Rector to 'present such hypotheses clearly only as a mathematician, who does not have to bother himself about the question whether these theories correspond to existing things or not' (p. 121). Kepler took the hint; but it was pretty evident that his youthful determination to 'ascribe to the Earth on physical, or if you prefer metaphysical, grounds' the motion then ascribed to the Sun could not be long silenced. In his next major astronomical work *The New Astronomy*, published in 1609 but virtually completed four years earlier, he removed any possibility of doubt as to his position by appending the famous expansion of the title to $\dot{\alpha}\iota\tau\iota o\lambda o\gamma\eta\tau\alpha$, *physica coelestis tradita commentariis de motibus stellae Martis* ('On celestial physics based on causal reasoning delivered by means of commentaries on the motions of the planet Mars'). This whole title was a manifesto of Kepler's scientific ('or if one prefers metaphysical') faith and the trumpet blast heralding the 'new' natural philosophy. It is a matter for regret that this is obscured by the title under which it is often known—*On the motions of the planet Mars*—which emphasises merely the source of the quantitative data on which the epoch-making theory was raised; the fact that the orbit of Mars renders more glaring the sterility of the traditional astronomy and physics is purely contingent. It is the double insistence—*physica and* $\dot{\alpha}\iota\tau\iota o\lambda o\gamma\eta\tau\alpha$—on the indissolubility of astronomy and physics that gave the work its almost unique character; that, and the author's laying bare of the almost superhuman intellectual and moral struggle by which it was brought to birth. Viewed thus from within, as it all too seldom has been, it stands out as unquestionably one of the major achievements of the human spirit.

Of the technical justification for this last claim it is unfortunately impossible to give more than a hint; its importance as a turning-point in the expressed relationship between Man and the Cosmos lies, however, not wholly, nor even perhaps mainly, in its creation of a 'new astronomy' as such, but rather in the nature of the sources from which Kepler's inspiration was drawn. Herein lies its high significance in a discussion entitled 'From magic to science'.

We may best appreciate Kepler's position by noting once again the titles of his two first major astronomical works—*The Cosmic Mystery* and *Celestial Physics*—and adding thereto *The Harmony of the World* (*Harmonices Mundi, libri v*) published in 1619 and considered by the author to be his supreme achievement. *Mystery— Physics—Harmony*: a strange trio—stranger still when Kepler's 'physics' is revealed in its fantastical garb—to be indissolubly linked with the 'origins of modern science'. For most (but fewer than heretofore?) modern scientists the link, if known at all, was only apparent, due, it was supposed, to the slightly off-beat nature of Kepler's mental processes. For such scientists the only thing that 'matters' is Kepler's clear enunciation of 'Kepler's Laws': (1) elliptical orbits, (2) velocities proportional to the *areas* swept out by the radii, (3) the squares of the periodic times of the planets proportional to the cubes of their mean distances from the Sun. In fact, so far are these from being 'clearly enunciated' that some wag once said that Newton's greatest discovery was Kepler's Laws! And by some unaccountable oversight the fundamental discovery—that the orbits of the planets are co-planar with the Sun—is never included. This discovery was fundamental in providing powerful evidence for what in Copernicus was no more than a fine careless rapture—that the Sun 'rules' the planets. This radical advance made by Kepler would have been sufficient to account for his saying that 'Copernicus was richer than he knew'.

Copernicus, it will be recalled (p. 121), based his faith in solar power on a number of classical traditions including the Hermetic: as to how this power was transmitted he made no specific allusion, but his reference to the Hermetic philosophy provides implicit evidence of what Kepler asserted explicitly; both astronomers were clearly under the spell of the Neoplatonic belief in the universal dissemination of cosmic power by emanation from the divine source; this was assimilated into an influential 'wing' of Christian philosophy through the writings of the so-called *pseudo*-Dionysius. The consequent alliance gave to certain magical yearnings and practices a 'respectability' they would not otherwise have enjoyed: the assessment of its importance for Renaissance science will be a major concern of the remainder of this chapter. But first we must return briefly to see how these 'unscientific' ideas provided the stimulus to Kepler's triumphant creation of the 'new' astronomy.

The personal source of Kepler's cosmic vision was less the Hermetic version than that of Nicholas of Cues—*divinus mihi Cusanus* he wrote in his first great work—in whose teaching the heady Neoplatonic speculation concerning the omnipresence of God was, as we might say, 'controlled' by the insistence to put all claims to natural knowledge to the test of measure and mathematical demonstration. But with this went the equal insistence on the *parity* of all situations in the universe, from which followed the rejection of the Aristotelian dichotomy between the 'perfect' celestial motions and the sub-lunary 'physics' of growth and decay; the same mathematical relations of power and harmony are to be sought throughout the Divine Creation. Proceeding in some ways more 'scientifically' (p. 147), in others with deeper physical insight, than Cusanus Kepler envisaged the Sun as 'ruling' the planets by rays of magnetic force; as the Sun rotated (as a few years later the observed motion of sunspots showed that it probably did) the rays, like the spokes of a wheel, swept the planets round. The dynamical analysis was wrong; the belief in the cosmic motive power of magnetism (despite Kepler's joy at Gilbert's 'verification', p. 168) was wrong; but the insistence on a *physical* link between Sun and planets to explain the 'mathematical' facts embodied in three of the four 'laws' was decisive for the whole future of science until the Relativity Principles involved a throw-back to that part of the teaching of Cusanus which Kepler was unable to accept.

In one respect however Kepler went further. Everyone knew that 'heavy' bodies fall towards the earth—their proper element whose centre was also the centre of the World; this universal characteristic was even then known as 'gravity'. Copernicus had already recognised that this behaviour had nothing to do with the element 'earth', or the planet 'Earth', henceforth to be regarded as a physically rather insignificant body hurtling through space at a hardly conceivable speed. It was, Copernicus thought, rather a 'natural striving' in the parts of all those bodies that display a spherical state. Kepler, on the other hand, insisted that gravity is a property of mutual attraction between separate *bodies*; if there were no restraining force 'the Earth would move up toward the Moon . . . and the Moon would come down to the Earth'. Likewise the oceanic tides represen-ted the 'rise and fall' of the waters towards the Moon. Was this attractive force constant? Yes and no. It was the same in all bodies of the same weight but fell off proportionately to the distance between the bodies.

What an astonishing series of 'near misses' arising out of a deep *feeling* for the omnipresence of cosmic 'powers'! But in natural philosophy a 'miss is as good as a mile'. No amount of attraction in a straight line can explain a continuous elliptic orbit; so the bogus 'magnetic vortex' had to be introduced. Also, despite Kepler's knowledge that *light-intensity* falls off according to the *square* of the distance, the same relationship was recognised as necessarily applying to any interplanetary forces only fifteen years after Kepler's death.

Finally as to 'harmony'. Whereas his first great work had demon-strated a fairly close (but probably fortuitous) agreement between the radii of the ('circular') planetary orbits and the relative sizes of 'nested' regular polyhedra, his last demonstrated a 'real' and far more important relation (p. 146) between the angular *velocities* (hence the eccentricities of the orbits) and the respective mean distances from the Sun. Thus not only does the solar 'power' become weaker with increased distance but to a degree expressible in a precise mathematical formula. And—wonder of wonders!—the ratios involved are similarly related to the accoustic harmonies, notably the diapason (octave), 2:1. For Kepler this was the moment of supreme triumph. His earlier researches, by showing the non-existence in the heavens of those 'perfect' mathematical objects— circle and uniform circular velocity—had in a sense detracted from Man's vision of the Divine plan; this was now restored, if not to its pristine simplicity, at least to a physical depth and mathematical subtlety that evoked from Kepler his most rhapsodical outburst:

Now that it is eighteen months since the dawn broke, three since the full light of day, and but a very few days since the clear Sun of a most marvellous vision began to shine, at last nothing restrains me; gladly I give myself up to a sacred frenzy, gladly scandalise mortals with the frank confession that I am spoiling the Egyptians of their sacred vessels in order to make of them a tabernacle to my God far from the borders of Egypt. If you pardon me, I rejoice; if you are angered, I can bear it. See then! I cast the die and write a book to be read either at the present time or by posterity; I care not. Let it wait a hundred years for a reader, as God himself waited six thousand for an observer.

After Galileo and Descartes had moved a long way from Kepler's misconceived dynamics it did not take astronomers very long to prove that, *if* the orbits can be assumed *circular*, then the 'Harmonic Law' implies an acceleration towards a fixed point varying inversely as the square of the distance. Kepler did not live to see this apparently inconsistent derivation. But, knowing from his deep study of Apollonios that the circle may properly be regarded as simply a *singular* plane section of a cone, his raptures would have had no bounds when the supreme genius of Newton removed the inconsistency by proving that, depending only on the initial conditions of the motion, the orbit can assume any one of the conic sections. Thus was the vision of the 'Divine Clockwork' completed.

Of the close involvement of Kepler's 'celestial harmonics' with the contemporary progress in the scientific study of music it will be proper to speak in the next chapter. Here finally we should take note of his recognition of 'static' harmonies throughout the sensual world in which we live. One charming opuscule, *On the Six-Cornered Snowflake* (1611), is in this respect outstanding. Here he notes the near-universality of regular shapes and symmetries expressible in simple numbers: a six-fold symmetry for the infinitely patterned snowflake, six-sided cells in the honeycomb, the great number of plants having *either* 5/10 (dicotyledons) *or* 3/6 (monocotyledons) floral organs or seed compartments; and much more. 'Economy of spatial packing', even mutual compression, as he shows by simple experiments, may be an essential causal factor, but a wholly satisfying explanation demands a recognition that within every object of regular and intelligible form there exists some 'power' akin to that which evidently informs every living thing. In his earlier works Kepler doesn't hesitate to speak of the 'soul' (*anima*) of a planet responding to the Divine power radiated by means of the Sun. Later (e.g. in *The New Astronomy*) he replaced the word *anima* by *vis*; but he never departed from a belief in what a modern writer has called a 'sacramental universe'.

These considerations have at length brought us back to the question of how far the decisive turning points in the scientific thought of the Renaissance were related to the tradition of 'natural magic' persisting throughout the sixteenth and into the seventeenth century. In particular the whole tenor of the *Six-Cornered Snowflake*, written as a kind of relaxation during the years in which the ultimate hypotheses of the *New Astronomy* were being 'forged in the teeth of irreducible and stubborn facts', is redolent of the animistic climate pervading the works of Paracelsus. Kepler in his worship of mathematics stood at the opposite pole to Paracelsus; nevertheless in developing a new insight into Appollonian conics he wrote: 'It is right that geometrical terms should serve us by way of analogy for I am especially fond of analogies as my most faithful instructors who are aware of all the secret powers (*arcana*) of nature.' Since the word *arcanum* is of course a classical Latin word for 'secret' (especially referring to sacred or mysterious things) too much should not be made of Kepler's choice of it in this context; but it was the term most favoured by Paracelsus for the hidden power it was the business of the alchemist to discover and apply in the relief of the disease related to it. Again, where Kepler sees mathematical forms as the 'signs' of natural relationships, Paracelsus dwells on the older and cruder doctrine of 'signatures' where a close *resemblance* is recognisable between, e.g., the liver and the liverwort —the plant indicated by God as a remedy for the diseased organ. Between Kepler and Paracelsus there may perhaps have been no close link—Kepler finds it necessary to condemn in the strongest terms the antics of many of the 'Paracelsists' of his own day—but we can hardly doubt that these two great workers in the cause of creating a 'new' approach to natural philosophy stood within the same tradition. This tradition we have already characterised in the phrase 'Man can do what he will'; there we were concerned more with performance in the 'arts'—'fine' and 'useful'—and (p. 77) in the art of government. Here we must try, however sketchily, to appreciate the accompanying 'theory'—especially that in respect of which Man, in his attempt to impose his will on the world of nature, allies himself with the powers hidden in that nature, or existing in a *supernatural* state, whether in harmony or in conflict with God's creation.

The most eloquent exposition of this 'policy', setting forth its justification and at the same time stressing the inherent dangers, was composed by Giovanni Pico, the young prince of the tiny city-state of Mirandola, as a *Prologue* to his 900 challenging theses he hoped might be debated with the 'reverend fathers' at that time controlling the destinies of the Roman Church. The young man's enthusiasm had, however, outrun his discretion: instead of a debate he got a papal condemnation, and but for the timely intervention of Lorenzo de' Medici might have come to a sticky end. The papal denunciation was finally withdrawn, but Pico died before publication of his theses or oration could be considered. The *Oration on the Dignity of Man*, as it came to be called, was first printed in 1498 in the *Opera* edited by Giovanni's nephew, Gianfrancesco; this edition contained also the elder Pico's famous *Disputations against Judicial [divinatricem] Astrology*.

The core of Pico's thesis concerning Man is twofold: that he was placed by God in a privileged position in the World, and that he is 'free'. This 'freedom' stems from the fact that Man is bound neither to a celestial condition like the angels, nor to the 'excrementitious and feculent parts of the lower world' like the remaining animals. He may thus make of himself what he will—to degenerate into a brute, or to soar by virtue of his spiritual nature into the upper realms of wisdom and light. And from his privileged position in this 'middle state' he can survey with greater ease and assurance than any other creature 'all that the great world inhabit'.

Among the 'arts' that Man may employ to achieve his highest good is that kind of magic known to the Greeks as μαγεια which must be clearly distinguished from that known as γοετια, for while the former is one in which the operator (*magus*) is as it were a servant and interpreter of nature, the latter kind is effected through the medium of demons. *Magia*, as Pico affirms more than once, is nothing but the 'consummation of natural philosophy'.

Pico's ability to read the Hebrew translations of the great 'Arabs' (especially Averroes) gave him a deeper respect for them than was possible to those of his contemporaries who were restricted to the Latin versions. He was thus able to call to the defence of his thesis an almost unbroken tradition from Moses through Pythagoras and Plato to St Augustine and the persistent strain of Neoplatonism in Christian mysticism. His familiarity with the Hermetic philosophy is revealed at the very beginning of the *Oration* where he quotes as the 'saying of Hermes Trismegistos, "A great miracle, O Asklepios, is Man" '. And not only in general terms does he claim the support of this tradition but in respect of the key to the secrets of nature which he claimed to have perfected. Plato and Aristotle, he reminded his hearers, were echoed by 'Avenzohar' (ibn Zuhr) in their belief that he who knows how to count (*numerare*) can know everything; among the medievals Roger Bacon had called mathematics the 'key' (*clavis*) to natural knowledge. Was then Pico, the humanist, as was his contemporary Leonardo da Vinci among the artist-craftsmen, an advocate of the 'mathematisation' of natural philosophy? Alas! No. For his 'mathematics' was not that of those he had claimed as supporters—he expressly warned his 'audience' against the kind used by merchants—but the numerology of the Hebrew *cabala* —a verbal tradition by which the adept was endowed with the means, so it was claimed, of calling upon the occult powers of heaven and earth by the utterance or display of words, letters, and numbers, of revealed potency and secret correspondences. Pico and his German contemporary, Iohann Reuchlin, must then be held responsible for the widespread grip of this baseless doctrine that during the following century often fogged the real issues of natural science. Nevertheless, men of the highest intellectual integrity such as Dee (p. 102) and Kepler were strongly drawn to it: Kepler in his controversy with Robert Fludd (p. 162) lived to regret his wasted hours; but can we be so sure that this to us futile motivation was not a necessary stage in the release of his cosmic imagination?

Was there then nothing really new in Pico's manifesto, and in his novel introduction of an Eastern occult 'art' into Western thought nothing but an obstacle to a more scientific way of thinking? Pico was indeed a child of his times, as was Roger Bacon before him and Copernicus later, in thinking that the 'truth-content' of his claims was in direct proportion to their antiquity; but probably never before had these ancient doctrines—and some further ones now to be brought forward—been so harmonised as to constitute a new attitude to Man and nature.

'Harmony' is here an operative word. Nearly two centuries before John Evelyn chose similar words for the motto of the newly formed Royal Society Pico staked his claim to impartiality thus: *in nullius verba iuratus me per omnes philosophiae magistros funderem*—he can spread himself throughout all the masters of philosophy, he will harmonise the apparently conflicting teaching of Plato and Aristotle, recognise the worth of the Arabs—even of Averroes whom Petrarch had called 'a frantic dog'.[1] This was no empty boast, nor was the outcome achieved by merely omitting the divergencies. Thus in his *Disputations against Judicial Astrology*, he adopted unconditionally the belief that 'apart from the influence of motion and light there resides in the heavenly regions [*caelestibus*] no power to them' and rejected the view (in which Roger Bacon had followed the Arabs) of a special medical virtue in the Moon. In this work another aspect of 'harmony' is developed: though it is unnecessary to look for 'occult causes' in the heavens to explain occult effects in terrestrial bodies, the power of heat and life-giving light are 'occult' in the sense that by their agency are the hidden purposes of God achieved. This 'harmony' ($\sigma\upsilon\mu\pi\alpha\theta\epsilon\iota\alpha$) of God and nature is stressed also near the end of the *Oration*, where having noted that the *magus* brings forth into the open the miracles 'concealed in the recesses of the

---

[1] The source is actually *Nullius addictus iurare in verba magistri* (Horace, *Epis*. 1.1.14 'Having bound myself to swear in the words of no master').

world', Pico emphasised that 'nothing moves one to religion and to the worship of God more than the diligent contemplation of the wonders of God'.

For the sake of clarity and brevity in exposition Pico has been given the central place: no one at this critical juncture showed so much insight or gave so balanced a sense of direction in which fruitful reassessment of nature might proceed. But it was his master, Marsilio Ficino, whose translation of the Hermetic works *first* focussed the attention of the thinkers of the later *quattrocento* on the question of magic and whose name and whose own works, rather than Pico's, are found most frequently cited among the writings of the *cinquecento*—and not only in Italy. This is understandable in view of the circumstances of the two men; but Ficino is of special importance, not only for having made available the supposed original semi-divine sources but also by virtue of his magical practices, in which the dividing line between the 'sacred' and the 'profane' is seen to be drawn much less clearly than he, later to become an ordained priest, would have cared to admit.

Born in 1433, and nurtured in the Aristotelian scholasticism that was still dominant in the universities, Ficino graduated in Medicine; but by the age of thirty he had become sufficiently prominent in the movement (given a great impetus by the 'invasion' of the 'Greeks' at the Council of Florence in 1439) towards Platonism for Cosimo de Medici to install him as the leader of the so-called 'Platonic Academy' at Careggi near Florence. The 'Platonism' that emerged in his original writings and was spread by his 'pupils', though based on his translations of the dialogues of Plato from Greek MSS, was that of a devout Christian and seen through the eyes of the so-called 'Neoplatonist' philosophers active in the Alexandrian world during the first four centuries of the Christian era. Though claiming Plato as their master these men could not escape the powerful influence of Aristotelian thought and the telling of fantastic outpourings of a civilisation at first invigorated but by then disintegrating under the impact of alien cultures.

An outstanding 'guide' to the cosmic mysteries in much of this many-sided culture was the collection of tracts now commonly referred to as the *Corpus Hermeticum*. Originally written in Greek, some of these were already approved by Tertullian about the beginning of the second century; the earliest extant *collection* is in a fourteenth-century MS. At Cosimo's request Ficino postponed his translation of the Platonic dialogues in order to provide fresh translations of some of these tracts. These translations were subsequently published under the misleading title of 'Poimander' which was the name of a participant in one of the dialogues; in a second dialogue the interlocutor is 'Asclepius'; in both it is 'Hermes Trismegistus' whose wisdom is being sought. This 'Thrice-great Hermes' was universally supposed to have been the Egyptian god, Thoth, whose nearest Greek equivalent was Hermes (Roman, Mercury) and was described by Tertullian as 'Master of all natural philosophers' (*physicorum*). His divine wisdom had, it was supposed, been culled by Greeks, probably Pythagoras, from Egyptian priestly sources to whom it had originally been revealed by the god himself. Plato, who was of course strongly influenced by the Pythagorean (number) mysticism, was the link with the Western world. 'Recessive' in the medieval world (perhaps owing to the Church's suspicion of 'magic' in any form) the 'wisdom of Hermes' seized the imagination of Renaissance society: in almost every one of its cultural activities its influence may be detected. The other main centre before 1500 was Paris, where Lefèvre d'Etaples made extensive translations and commentaries. Its influence is evident in the *Foure Hymnes* of Spenser, in the poems of Philip Sidney, and in the strange happenings in the house of John Dee at Mortlake. But it was in England that the Swiss scholar, Isaac Casaubon, showed by textual criticism that the *Corpus* could not possibly have been written before the second century of the Christian era, thus reversing the traditional myth: instead of being a miraculous anticipation of Christianity it was inevitably to some degree its product. Also, so far from having been the *source* of Plato's wisdom it was rather a vision of Plato's own thought largely transfigured through Aristotelian and Stoic eyes.

The element in Ficino's magic that Pico, as respectfully towards his master as might be, condemned was that in which the former hoped to draw spiritual power from the 'world-soul'—an essential assumption in Neoplatonic philosophy. This could be achieved, so it was generally believed, by taking into the physical body such things—wine, the 'quintessences' of roses and gold, spices such as cinnamon—in which the 'world-soul' might be expected to inhere; the potency of this practice could be enhanced by listening to, or actually singing, suitable music such as the orphic hymns. Most potent of all, however, was the 'attraction' of individual spiritual powers into images or carefully composed inscriptions whose recital would add the more powerful element of *sound* to that of sight which is limited to static forms. To the 'milder' ritual there was no ecclesiastical objection; against the latter, which involved Man's *spiritual* nature, there was a definite veto—for the plausible reason that the operator would be unable to distinguish the cooperation of evil 'demons' from that of the good.

Ficino was no more—rather less—a 'scientist' than Pico; his influence among those who, though still remote from the 'scientific revolution', were moving more closely to its sceptical attitudes, may have been at least as important in the opposition it engendered in them as in the wide acceptance, not always explicit, of his basic assumptions as to the relations of Nature, Man, and God. This pervasive influence, far beyond his 'ivory tower' at Careggi, is hardly surprising in view of his many-faceted character. To an Age of Reason as to an Age of System his teaching appeared as a hotch-potch of former views—mostly erroneous—exemplified in a life without settled aim or judgment, as such he gave eloquent expression to the intellectual and moral confusion of an age in which every aspect of human nature was being called in question with a degree of violence and arrogance comparable to our own. His sense of style, learning, and piety gave a spurious air of academic respecta-bility to opinions and practices that might easily have undermined the order and rationality by which the medievals had created a high civilisation on the ruins of Imperial Rome. Fortunately among those who consciously or otherwise saw in Ficino's world-view a redemptive power there were some who were ready to reject un-compromisingly those aspects of it that they believed to be vicious. Pico, as we have seen, put the problem of astral influence in a new light—a correction that Ficino was not altogether unwilling to accept. How much more that young genius might have achieved we can barely conjecture. Considerations of space allow of reference to no more than three later figures related to the 'magical' tradition; 'eminent contemporaries' of each other, but two of whom—Heinrich Cornelius Agrippa of Nettesheim and Pietro Pomponazzi of Mantua —may probably have influenced the youngest, Paracelsus.

Pomponazzi was born in the year (1462) in which the Careggi 'Academy' was founded, the 'Platonism' of which was thus an established influence before he was old enough to react against it. In fact the development of his thought and that of Ficino were largely complementary; this was a consequence of the younger man's strictly 'academic' career in contrast to the 'privileged' and perhaps rather *dilettanti* circle of Careggi. Pomponazzi also moved almost wholly in a 'circle', but it was the extended academic circle of Padua, Bologna, and Ferrara; in all of which, though mainly in the first, he had to prove himself at least the equal of a professor who, with more justice than is now usual with this much abused term, could be called his 'opposite number'. The opposition represented in these disputations was between the traditional Aristotelianism of the 'Commentator' (Averroes) and that of Aristotle himself, by then made possible through the more definitive texts translated by 'human-ist' scholars from the original Greek. This brief glance at Pompon-azzi's academic situation was necessary to pinpoint his peculiar and subtle contribution to the assessment of 'natural magic' as 'natural philosophy'.

Throughout the *quattrocento* at Padua the orthodox *philosophical* view of the human soul was that of Averroes, namely that it was only as it were an influx of the world-soul into the matter of the individual human body. The profound *theological* consequences of this view need not concern us; but in rejecting this really Neo-platonic reading of Aristotle Pomponazzi developed an alternative that in a fundamental though not obvious way reveals a 'scientific' temper hardly paralleled at that time.

At a superficial level the work of Pomponazzi *On the Immortality of the Soul* is retrograde; for although he accepted Pico's hermetic interpretation of Man as a great marvel with power to make of himself whatever else there is in the macrocosm, yet at the same time he restored the paramountcy of the heavenly bodies as such. They alone determine and effectuate the actions and passions of men; 'nor is it strange that such things can be shadowed forth by the heavenly bodies since they are animated by a most noble soul and generate and govern all things below'. We may however see in this restoration a compensatory advantage, since it removes the necessity for assuming the existence of an innumerable host of *para*-spiritual powers including the personal 'genius' or 'familiar demon': 'If we can do without that multiplication of demons and genii it seems superfluous to assume them; besides the fact that it is also contrary to reason.' This is good Occamist stuff: which is the highest praise that could be given by 'hard' scientists! The implications for natural philosophy are more fully explored in his book *On the Causes of Natural Marvels [admirandorum] and On Incantations* where he maintained that no natural event is brought about by the action of celestial powers except mediately through the heavenly bodies. Hence nothing happens arbitrarily or at the whim of any supernatural agency; all is bound together in a system whose 'rules' can be read in the skies. We may smile at the artificiality of the supposed mechanism; but the system is remarkably like that of the supposed 'laws' of nature that 'can be inferred from phenomena and admit of no exception' (Isaac Newton—somewhat condensed). The consequences for theology drawn by Pomponazzi were not as might have been expected that the immortality of the soul is a pure myth; he preferred to 'keep the saner view', namely, 'that the question of the immortality of the soul is a neutral problem like that of the eternity of the world. For it seems to me that no natural reasons can be brought forth proving that the soul is immortal and still less that the soul is mortal.'

It is easy to see where Paracelsus might have got his 'stellar pathology' (p. 88) from, and the possible basis of that far more rigorously derived *Ethic* of the 'celebrated atheist' and 'God-intoxicated' Jew, Baruch d'Espinosa. A more distant echo may perhaps be heard in Kant's conclusion that it may be necessary to 'set aside reason to make room for faith'. Even 'Darwin's Bulldog', Thomas Huxley, when not engaged in bishop-baiting, might have seen the relevance of this scholastic-humanist enquiry to his own 'agnosticism'. The epithet 'scholastic-humanist' is here no contradiction in terms, but is used deliberately to highlight the artificiality of many of the neat labels beloved of academics. Not the least significance of Pomponazzi is his effective combination of scholastic philosophical penetration and rigour with the insight into the original texts made possible by the linguistic precision of 'humanist' grammarians and philologists; likewise his recognition of the (Neo) 'Platonism' at the heart of Averroistic 'Aristotelianism'. Here then was another 'Conciliator' like Pietro d'Abano (p. 92), a 'Princeps Concordiae' like Pico; a thinker from whom all the ages might learn the many ways leading to the unrealisable ideal of 'ultimate' truth. As such, living as he did in a Renaissance society, he survived to see his works committed to the flames; in a region and at an epoch in which ecclesiastical power had been less tamed it might not have been only his works that the flames devoured.

It was not the *existence* of 'marvels' that Pomponazzi called in question, only their order and relationship to the celestial and terrestrial regions. We may, if we like, object that those who are not 'marvels' that occur 'according to law'. This objection is well taken; but linguistic habits are not changed in a generation: the close affinity between Paracelsus' theory set forth in a highly technical 'magical' terminology and Pomponazzi's more rational *schema* affords sufficient example. In any case magic, like sex in our time, was a saleable commodity unlikely to be promoted solely by critical and rational means. Of the hordes of 'adepts' claiming to convert base metal to gold, to distil 'arcana' productive of virility and long life, or to guide the destinies of princes by converse with astral demons, and the like, we can but take note; like the poor they are always with us, even—nay, more so—in a 'Scientific Age'. But to the academic theorist and medical magician we must now add, as highly influential in Renaissance society, a third type combining to some extent the characteristics of the first two with a passionate concern for the application of the 'higher learning' to the problems of social Man. This was H. C. Agrippa of Nettesheim.

Of this highly complex figure it is possible only to give the barest sketch if only to correct the very partial and even false references that are still current. In training and practice he was the 'typical' Renaissance Man: he took all knowledge—Law, Medicine, Theology, Philosophy—for his province, spicing the more orthodox university courses with magic, cabala and cryptography. Academically nurtured mainly at the University of Cologne, he lectured at several centres in Northern Italy, practised as municipal physician in Switzerland, acted as medical adviser to the Queen-Mother in France, went in embassy to London (where he took the opportunity to study with John Colet), and finally died, embittered and in extreme poverty, in France. But he was by no means the charlatan he has been made out to be; his learning was praised by Erasmus and he was ahead— far ahead—of his times in publishing a widely read book on the superiority of the female sex. Though rushing into magical practices —learnt mainly from the *Picatrix*, a medieval Arabic (per)version of the Hermetic works—where Ficino, whom he quoted at length, feared to tread, he did so in the conviction that divine power would always save the true believer from the machinations of evil demons. One major paradox in his life has been explained only rather unconvincingly by Agrippa himself. This occurs in his dedication to Hermann, Archbishop of Cologne, of the first edition of *The Occult Philosophy* dated 1531 but probably first published in 1533. The paradox lies in the fact that the views set forth in the book had been categorically renounced in his recently published (1530) *Uncertainty and Vanity of Sciences and Arts and an outstanding Declamation of the Word of God*. The 'explanation' he gives of the publication of the *Occult Philosophy* was, he alleged, the necessity for an accessible authentic version of a youthful work, previously circulated in manuscript (*c.* 1510), to replace those corrupted by the hands of others. There is no little evidence to confirm the need for such an extravagant procedure; we cannot rule out the possibility that the *Uncertainty* was a kite flown both to advertise the coming *Occult Philosophy* and to anticipate any ecclesiastical onslaught that the latter might have prompted. Against this conjecture must be balanced the character of the author as revealed in his life—restless, impulsive, devout and arrogant by turns, now walking with kings, and again seeking any means to stave off creditors: more Faustian in fact than the Dr Johann Faust named in the first printed version of the legend, and far more fully authenticated. His importance lies not in any solid contribution to knowledge nor in any novel interpretation of the universe, but rather as perhaps the first man of noble origin and academic learning who became a focus for the growing obsession with the so-called 'black' magic. For many of Agrippa's contemporaries, even some of undoubted integrity, this practice was one by which 'men can do all things if they will'. But it carried with it grave risks of spiritual disintegration. Though the British Museum lists nearly twenty editions and translations of *The Uncertainty* and only five of the *Occult Philosophy*—one bearing the autograph (amanuensis?) of Archbishop Cranmer—Agrippa's reputation has not surprisingly rested on the infamous practices hinted at in the latter rather than on the disillusionment and recantation voiced in the former.

1 2

'THE GREAT INSTAURATION'

The title is that of Francis Bacon's most influential work but it may equally stand for a turning point in history. Of course, Bacon was not, as Thomas Sprat claimed, 'the one great man who had the true Imagination of the whole extent of this Enterprize'; this was the work of many hands, to some of whom reference has already been made. To most of his contemporaries, and to many of his successors, however, there was no one more representative of the Great Instauration than Bacon. For this reason we are perhaps justified in making a virtue of the necessity imposed by considerations of space of taking him as the central figure into whose works many of the currents of Renaissance thought flowed and from which a vision of the future could be most extensively and clearly envisaged. At the same time the fact that the extent of Bacon's contribution to the emergent 'new philosophy' and its contemporary society is still a matter of even violent controversy makes it essential to hear his own version of what he was about.

In setting forth the nature of the enterprise to which *The Great Instauration* (*Instauratio magna*) was to show the way Bacon claimed that 'the matter in hand is no mere felicity of speculation but the real business and fortune of the human race and all power of operation. For Man is but the servant and interpreter of nature; what he does and what he knows is only what he has observed of nature's order in fact or thought: beyond this he knows nothing and can do nothing. For the chain of causes cannot by any force be loosed or broken nor can nature be commanded except by being obeyed. And so those twin objects, human knowledge and human power, do really meet in one; and it is in ignorance of causes that operation fails.' As such, Bacon's work ought to be judged: not as a 'primer of the scientific method' nor as an exposition of inductive logic; but, in Professor Benjamin Farrington's words, 'as a blueprint for a new world'.

From the date (1620) of its appearance *The Great Instauration* might be held to lie beyond the limits we have agreed to set upon the Renaissance. It was, however, an attempt to systematise an ideal clearly envisaged and expressed by Bacon some years before the end of the previous century, and to incorporate a somewhat toned-down diatribe against those 'corrupters of philosophy', Plato and Aristotle. Neither of these relatively youthful effusions was ever published, but in 1605 appeared a rich if somewhat adipose review of the 'deficiences' of existing knowledge, in the course of which are to be found many very acute observations on the means of making them good. Within this *Advancement of Learning* may be found all, or nearly all, the features that were later to be included in the *Great Instauration*—the sterility of syllogistic logic when applied to ill-conceived premises, the primacy of experience enlarged and controlled by experiment, the revolutionary power of instruments such as the magnetic compass, the necessity for the collaboration of men of various talents, and—what critics have been slow to credit him with—the recognition that 'to conclude upon an enumeration of particulars, without instance contradictory, is no conclusion but a conjecture': all these, and much more, especially the sentiment, everywhere implicit in his works but here made explicit, character-ising knowledge as 'a rich storehouse for the glory of the Creator and the relief of man's estate'. The *duality* of this aim cannot be too much stressed; nowhere is it more beautifully expressed than in the lines very near the beginning of the *Advancement*: 'Let no man . . . think or maintain that a man can search too far or be too well studied in the book of God's word, or in the book of God's works, divinity or [natural] philosophy; but rather let men endeavour an endless progression in both: only let men beware that they apply both to charity, and not to swelling; to use and not to ostentation; and again, that they do not unwisely mingle or confound these learnings together.' The warning was, at that moment of history, no less important than the aim. Throughout the seventeenth century the dogmas of rival religions, far more than of rival 'philosophies', engaged men's intellects as well as their emotions; and this went also for men like Napier, Boyle, and Newton, to whom persecution and war against rival sectaries would have been abhorrent. It was therefore essential for the progress of science to 'free it from the passions of sects', later noted by Bishop Sprat as among the ideals of the Royal Society. Galileo, a man of very different temper from Bacon, was saying much the same thing, though hardly acting up to his precept. Today, when the Christian churches are seeking to emulate the humility of their Founder, and when probably the majority of men of science accept the innate element of uncertainty in all their pronouncements, Bacon's regulated dichotomy no longer satisfies the questing spirit. Those seekers for whom the message of the late Père Teilhard de Chardin is a rallying point would agree unreservedly with Bacon's call for an 'endless progression in both' while accepting his 'warning' only with considerable reservations.

The *Advancement* was no more than a brilliant reconnaissance; the campaign was to be conducted in a series of volumes of which the *Great Instauration* of 1620 was intended to be the first. Essentially, this work contains a challenging 'Plan of the Work' (*Distributio Operis*), the major work known as the *Novum Organum,* and a number of promises, very few of which were ever fulfilled. Of these the only one of weight was the greatly enlarged version of the *Advancement* written in Latin under the title *De Augmentis Scientiarum*.

The *Novum Organum* (aimed at replacing the *Organon* or logical works of Aristotle) is commonly assumed to be the definitive exposition of Bacon's 'method'. Viewed in this light it must be confessed that it embraces both too little and too much: too little, since it contains no *systematic* exposition of Bacon's 'art of discovery'; too much, in that the only exemplification of what Bacon had in mind consists of twenty-seven 'prerogative instances' in which much acute observation is smothered in a mass of detail whose 'organisation' under a highly verbalised system of categories could certainly justify William Harvey's disdainful remark (if we can trust the garrulous John Aubrey) that Bacon 'wrote philosophy like a Lord Chancellor'. Setting aside this monstrous growth (the second of the two 'books') the *Novum Organum* contains in the *Aphorisms* of the First Book a wealth of brilliant singular insights into the scientific *temper* together with a delineation of 'the Four Idols that beset the human mind'. The names of these 'Idols'—of the Cave (individual human temperament and upbringing); of the Tribe ('human nature' in general); of the Market (the tyranny of words and clichés); and of the Theatre (the elaborate 'shows' contrived by system-makers) —illustrate, as does the strangely neglected *Wisdom of the Ancients* (1609), Bacon's love of teaching by myth and parable. In doing so he was only applying to natural philosophy a custom already widely adopted by contemporary English poets. But whereas the poets employed these myths mainly as literary conceits to give colour and vivacity to ancient wisdom there is evidence that Bacon recognised in them a medium whereby to insinuate his new and revolutionary ideas under the cloak of familiar literary conventions.

Looking back on this very brief sketch of Bacon's manifesto for his own times we can agree that he was no 'scientist'—while he was bemoaning the deficiencies others were making them good on a scale previously unparalleled. Nor was he a philosopher in the great tradition. In hardly any respect was he an innovator: the effective element in his 'induction' goes back through Paduan Aristotelianism to the *Posterior Analytics* of the Master himself (p. 92); his insistence on 'power and operation' through Pico della Mirandola to the Hermetic writings 'Pimander' and 'Asclepius', where also is to be found the affirmation of the means to power, namely, in obedience to the law-like order of nature. His sceptical approach to the existing natural knowledge had already been anticipated by Michel de Montaigne, who had a very low opinion of contemporary Medicine. In his insistence on the necessity for 'philosophers' to be also 'operators' the young Francis Bacon was only the latest of a long line of scholars going back through Agricola, Vesalius, Juan Luis Vives to Dürer and the artists of the *quattrocento*. It is now becoming more and more apparent that Bacon's outstanding achievement was his recognition that though the hermetic-magical strain, which had been so powerful a catalyst for Renaissance thought, had a superficial resemblance to his own design there was a fundamental difference of aim that necessitates a no less radical modification of method. 'Men can do all things if they will' provided that what they will is to achieve spiritual harmony with the cosmos. If their primary aim, as it was his, is the 'relief of man's estate' then they can *not* do '*all* things' but must limit their desires to the 'effecting of all things possible' (*New Atlantis*). The dedicated search of the Magus for 'instant' power is wholly irrelevant: in rejecting ceremonies to achieve natural ends Bacon was nearer to Paracelsus than he knew. Nevertheless, to the magical tradition he probably owed his reform of the empirical operation, namely that it is not enough merely to 'ape' (see p. 162) nature; her secrets must be wrested from her by contriving situations in which she is *forced* to reply.

Many less fundamental parallels could be cited. In one respect Bacon's insight so far exceeded that of his contemporaries that only in our time was his vision translated into fact. In the posthumously published but effectively complete fable, *The New Atlantis*, the 'enlargement of the bounds of human empire' was to be the 'end of our foundation'. What follows reads like a prospectus for the setting up of cooperative research associations among which may for instance be discerned the vision of the Low Temperature Research Association (founded at Cambridge) and the Scottish Hydro-electric Board. The *New Atlantis* was published in 1627. On 15 July 1662 letters patent were promulgated in which Charles II announced *Diumultumque apud nos statuimus, ut imperii fines sic etiam artes atque scientias promovere* ('to *extend the bounds of empire* as also the arts and sciences themselves'); this phraseology, that otherwise seems hardly relevant, is unlikely to have been a coincidence. Bacon has been described as one 'who rang the bell to call the wits together': the 'wits' to whom the king gave their first charter 'studied to make it not only an enterprise of one season . . . but a business of time; a steddy, a lasting, a popular and uninterrupted work'. Not only was this aim, as described by their first historian (or perhaps apologist), Thomas Sprat, characteristically Baconian but also the machinery by which it was implemented. The Royal Society proceeded to set up a number of committees, each under the supervision of an elected and suitably qualified 'curator', to enquire into and report on various 'arts' such as agriculture and ordnance. In addition there was a 'Curator by Office', Robert Hooke, who was a paid (when funds permitted!) official whose function was to bring forward experiments of a more fundamental character; for assistance in the assembly of the apparatus he could call upon the skilled mechanic, Richard Shortgrave, the paid 'Operator'.

During the years in which Lord High Chancellor Bacon was 'legislating' for the coming 'Scientific Revolution' Kepler was demonstrating that although the Hermetic sage might be an inspiring guide to the door of the castle of celestial knowledge, only the exercise of reason on the 'irreducible and stubborn facts' could force the door and reveal the true celestial relations. The sense of his conviction emerged in a famous controversy in which the English physician and admired friend of William Harvey, Robert Fludd, attacked Kepler in print for having in an appendix to his *Harmony of the World* (p. 145) made light of the 'sacred and celestial harmonies' that Fludd claimed to have established. The ensuing controversy sheds so revealing a light on this period of transition and on the leading figures drawn into it that a study of the main issues will repay the time spent on apparently fruitless sophistries.

In 1617 appeared the first of several huge volumes composed by Fludd under the general title *A metaphysical, physical, and technical account [historia] of each of the Two Worlds, the greater and the less.* The 'two worlds' were of course the at that time familiar 'macrocosm' —the whole Divine Creation—and 'microcosm'—Man who was held to mirror the former. The programme set out on the title-page shows that the completed work was to deal with all the 'arts' that may be brought to fruition in the macrocosm by the 'Ape of Nature', Man. Fludd's aim, broadly speaking, appears to be an 'instauration' on Baconian lines: men could 'do all things if they will' by 'aping', that is, imitating, nature. But even a rather superficial examination of the text and somewhat fantastic illustrations reveals that this consummation was to be attained through the recovery of the ancient wisdom by submission to the World-Soul, among whose instruments were alchemy, astrology, and above all, the celestial harmonies and symbolic numbers. It might therefore rather be regarded as an *anti*-instauration, pinning its faith on just those prevailing modes of thought that Bacon was at pains to repudiate.

Though there are indications in the *Novum Organum* and elsewhere that Bacon was aware of Fludd's cabbalistic notions (which of course he denounced) he did not refer to him by name. This can hardly have been due to a belief that a reasoned confutation was hardly worth the trouble, since Fludd's reputation for learning stood very high, and he was known to be the 'ambassador' in England for the Rosicrucian movement that from 1615 was sweeping over Europe. How far Kepler recognised a serious menace in this retrogressive trend it is hard to say; but in the *Appendix* already referred to he was at pains to distinguish bluntly between Fludd's bogus and his own genuine mathematical flights, as also to clarify the nature of the celestial vision that each of them believed he had achieved. The world-view of both was based on a sure conviction of the transcendent nature of the harmonic ratios (p. 53) in the pattern of which God had 'ordered all things in number and measure'. But whereas Fludd had, on the analogy of the monochord, constructed a completely imaginary universe in which the central position of the Sun, and of the heart in animals, was 'explained' by the unison of the two octaves of the instrument when divided at its centre, Kepler claimed that the harmonic characteristics he had discovered in the orbits of the planets were physical realities.

To the rising generation of 'scientists' this style of refutation was not going far enough. In 1616 one of them, Marin Mersenne, Minorite friar and friend of Descartes, published a large work, *Questions on Genesis*, in which he launched an almost scurrilous attack against Fludd who in due course hurled back insult for insult. This exhibition of intolerance appears to us all the more regrettable in view of the claim of Mersenne to be accounted one of the pioneers of experimental quantitative physics—a claim perhaps stronger than Galileo's: by his careful measurements and subsequent calculations of a sounding monochord he had gone a long way towards establishing the fundamental laws of acoustics as they have been known and applied ever since (but see p. 165). Instead of replying himself to Fludd's counterblast, he induced his friend Pierre Gassendi to belabour Fludd. More tolerant of rival views Gassendi adopted a less abusive tone towards Fludd but at the same time made the mistake of rejecting the new theory of the circulation of the blood to which his attention had been drawn by Mersenne. The verdict of history would have been in his favour had he merely rejected Fludd's bogus *reasons* for accepting his friend

Harvey's great discovery. But he rejected the whole demonstration on the grounds that he had himself seen the interventricular pores on the non-existence of which Harvey's demonstration turned. No better illustration could be adduced of the insufficiency of regarding mathematics as the only 'language in which the book of nature is written'. In respect of the problem set by the mode of action of the heart in the living body the 'hard-headed' mathematically minded Gassendi, as also Mersenne, maintained the wholly erroneous Galenic account; only the eye trained to read the language of organic form and its correlated function could see what was 'really there'. Fludd's mathematical fictions were here irrelevant; and Harvey of course used no mathematics except the knowledge that a thousand times half an ounce of blood is a lot more than nine pounds!

Though Kepler took no part in this particular aspect of the controversy, his final remarks on the relevance of the celestial harmonies reveal an uncanny genius for picking his way through the thicket of confusion characteristic of this time of transition. Though the inspiration of the ancient cosmic mysteries never left him, they served him only as a guide to the interpretation of the observed facts from which he started. Because Plato's *tetraktys* (p. 54) excluded the third and sixth harmonics as just consonances despite the evidence of the human ear, Kepler came to see that Pythagorean hypostasy of pure numbers was unjustified. It was not true that 'things are numbers'; even less were numbers things. The power of numbers (as he learnt from Aristotle) lies in the fact that, being abstracted from things, they may serve to represent them in the manageable relationships of arithmetic and especially geometry. This being so it was wrong, he thought, to discuss the possibility that, though inaudible to men, the celestial harmonies might be heard by the denizens of the superior hierarchies; sound was a *physical* property, hence improperly associated with *celestial* creatures. What then, we might ask, was left of the harmonic relations of the planets in which he took so much delight? His reply indicates a depth of understanding which even today natural philosophers might find enlightening: his periodic law—a generalisation of observed facts—proves that the orbital velocities of the planets stand in the same numerical proportion as do those harmonic ratios that are pleasing to our ears, thus revealing their common origin in a Divine mathematical plan. To paraphrase one of his most pregnant sayings: 'I play with symbols, but I don't forget that I am playing; for no deep secret of natural philosophy can be derived from symbols unless these are already associated with ideas based on reason.' In the cruder but equally significant phrase of our computer-scientists: 'Rubbish in, rubbish out.'

It was not by mere chance that music played so important a part—

how important has been stressed only comparatively recently—in the transition from the mystical view of the cosmos dominant in the Renaissance to that characteristic of the New Instauration. Of the branches of pre-electronic 'classical' physics, mechanics, optics, and acoustics had been studied critically and effectively at least from the time of Plato; the reason is not far to seek; they were amenable to mathematical treatment. Music, it will be recalled, was a part of the medieval quadrivium (p. 20). Just as mechanics throughout the sixteenth century and optics towards its end received a stimulus from the need to solve practical problems so the notable enthusiasm for 'courtly' music (evidenced in the paintings of many Italian artists) was followed by a renewed and sustained interest in musical theory. To the form of staff notation by then well established was added an alternative (tablature) providing a space for each string of the lute. Towards the end of the fifteenth century the number of printed editions of the classical textbook by Boethius (fifth century) shows that even the universities were being caught up in the vogue. To extend the harmonic range of stringed instruments, demands unlike the previously dominant human voice, a thorough under-standing of the relation between the physical characteristics of the string and the pitch it emits when plucked. A change in attitude to this problem is revealed in the following challenge: 'I wish to point out two false opinions of which men have been persuaded by various writings and which I myself shared until I ascertained the truth by means of experiment, the teacher of all things.' The words are those of Vincenzo Galilei, father of Galileo and an accomplished lutanist, from his printed *Discourse* of 1589 in which he demonstrated against his master, Gioseffo Zarlino, that only actual trial and *subsequent measurement* could establish the facts of just consonance. An unpublished manuscript of later date shows that the experiment described (*Discorsi* 1638) by his more famous son had been carried out at a time when father and son were living to-gether. By 1638 Mersenne had established the laws in a more systematic manner. The persistence of this association of musical theory with mathematics and natural knowledge had been given striking testimony when the forward-looking and pragmatic Sir Thomas Gresham included in the scheme for his college (p. 115) a Professor of Music to rank equally with those of Geometry, Astronomy, Law, Rhetoric, and Medicine.

That Fludd's kind of mathematics was a perversion that Kepler, Mersenne and Gassendi did well to attack there can be no doubt; his devotion to chemistry is less easily disposed of. Not that he made any positive contribution to knowledge even in that sphere; but by his insistence that only alchemy could reveal the 'hidden powers' in nature, and that this enquiry, though guided by the 'Divine

Light', must be based on experiment he was in effect, though neither of them might have cared to admit it, allying himself with the aspirations of Francis Bacon. His influence was enhanced by his high professional standing in the College of Physicians and by his hostility to Aristotelianism equal to Bacon's. In almost every respect his approach to Medicine was that of the progressive Paracelsians who, whatever their merits, were increasing in numbers in England: Sir Theodore Turquet de Mayerne, whose clinical history of Prince Henry's fatal attack of typhoid is commended by Sir Geoffrey Keynes as a model of its kind, had been virulently attacked by the *Faculté* in Paris for employing drugs based on metallic preparations.

The extent of the debt owed by Robert Boyle's science of 'chymistry' to alchemy is difficult to determine; but however wild may have been Fludd's conception of the 'aerial saltpetre' and the like it was a recognition of a pervasive and variously effective 'power' (characterised as oxygen more than a century later) such as the purely mechanical approach of Mersenne could never have achieved. Kepler, as usual, knew better; in the *Six-Cornered Snowflake* (p. 148) he was prepared to hand over to 'chemists and botanists' the explanation of the *vis formatrix* (his term for what has been regarded as something akin to Paracelsus' *iliaster* or *archeus*) which he saw everywhere at work in the regular floral and crystal patterns, and eminently so in the infinite variations on a hexagonal theme in the snowflake; all these his mathematical analogies had most beautifully correlated but were unable to suggest any reason why they should have come into being. Kepler, like Leibniz after the microscope had revealed organisation at a level far below that of the 'microcosm', recognised that here were problems of a different *order* from those posed by planets and cannon balls and demanding for their solution entirely different modes of thought. Inadequately conceived as they were, Fludd's hierarchical levels of being may not have been without any constructive value. In addition they probably gave some degree of cohesion to the protests levelled at the universities for their failure to provide instruction in 'chymistry'. But they also gave encouragement to the spread of alchemy and astrology that in England and Scotland proliferated to a previously unparalleled extent. This mainly uncritical aftermath of a once creative force reached its climax at the same time as the so-called 'Scientific Revolution': its *doyen* was the otherwise critical scholar, Elias Ashmole, an original Fellow of the Royal Society.

The extreme position sometimes adopted—that Bacon's whole enterprise was an irrelevance—perhaps, like the similar assessment of the influence of the humanists of the *quattrocento*, even an obstacle was based largely on his failure to appreciate the epoch-making discoveries of his contemporaries, notably Gilbert, Galileo, Kepler and Harvey. There is indeed no doubt that these four provided the essential basis for all future progress in scientific discovery. But to reject Bacon's testimony on the ground of his ignorance or even adverse criticism of the pioneers is quite another matter. Harvey's book didn't appear until after Bacon's death; previously exposition had been restricted to the Lumleian Lectures to the College of Physicians and at a date later than used to be accepted. Similar considerations apply to Galileo's fundamental reform of mechanics but not to his telescopic discoveries, which, as well as Kepler's innovation in planetary theory, were well known in England nearly a decade before *The New Instauration*. Bacon does indeed treat the former rather cavalierly, but emphasises the importance of the telescope as an instrument of discovery. It is not inconceivable that he regarded this greatly extended knowledge of the celestial regions as irrelevant to his purpose; his failure even to mention Kepler might likewise be explained by his inadequate appreciation of the power of mathematics, though he was far from rejecting mathematics as an aid to discovery. Of Gilbert's book *On the Magnet,* as also of the posthumously published work *On our Sublunary World* known to have been available in manuscript in Bacon's youth, he had a pretty thorough knowledge. His appreciation of the man and his work sheds further light not only on Bacon but on the peculiar dangers involved in estimating the degree of 'revolution' to be attributed to that cultural epoch.

Bacon's references to Gilbert are widely spread and are more often adversely critical than eulogistic. But what Bacon is objecting to is not Gilbert's 'science' but his 'philosophy', if we may make use of a modern distinction. Of his observations Bacon says that they were 'collected with great sagacity and industry'. Where Gilbert went wrong, Bacon asserts in more than one context, was in erecting on so narrow a base (magnetism) so large a 'philosophy': not without justice he regarded this as an instance of the *Idols of the Cave* (p. 160).

The conventional assessment of Gilbert takes the title of his work quite literally, noting especially the last phrase 'a new physiology [p. 140] demonstrated by very many arguments and experiments'. So far as it goes this assessment is unexceptionable: in no other work of the period is such a varied and relevant range of experiments deployed; it came as near as possible to being the model of the 'experimental philosophy'. But, as Zilsel pointed out in 1941, only about forty per cent of the text is concerned with physical magnetism as such (with a very brief but important section on the distinct nature of the electric 'effluvium'), the remainder ranging widely over its real and imaginary applications, notably navigation. Towards the end of the book, however, comes a hint of a neglected complementary approach to the whole enterprise: the magnetic force does not act by attraction of a single body for another but rather by a mutual rushing together and harmony (*concordantia*) for which reason Gilbert preferred the term *coitio*. There is here nothing *necessarily* inconsistent with the 'hard line' in natural philosophy that Mersenne was later to campaign for: the analogies might be merely rhetorical flourishes such as were commonly used at that time. But Gilbert went further: *Vis magnetica animata est*. Aristotle, he noted, had erred in restricting an *anima* to the celestial bodies; Plato and many others—Hermes, Zoroaster and Orpheus—on the contrary had acknowledged that everything (*universalis*) is animated; and for himself Gilbert believed that everything, including motion, was governed and maintained by *anima*. The word *anima* is here left untranslated to emphasise that what is in question can not be the 'vegetative' and 'sentient' souls (cf. p. 169) attributed by Aristotle to living creatures, nor clearly the 'rational' by which he distinguished Man; it seems rather to have been the 'love' for 'God' that maintained the eternal motion of the 'stars'. Leonardo's *spirito* (p. 52) manifested in *forza* seems still to have been influencing the thoughts of these pioneers of the 'experimental method'. The quoted passages in which Gilbert rejects what we should regard as the sound scientific common-sense element in Aristotle's thought only to accept the fantastical in Plato's shows once again the somewhat spurious character of the judgment that the 'Scientific Revolution' rested upon a return to 'Platonism' and a rejection of Aristotle. Finally Gilbert appeals to apocryphal 'authorities'—Hermes, Zoroaster, Orpheus—reveals the persistence in Gilbert's mind of the 'myth' of the Ancient Wisdom (*prisca theologia*, p. 121) which we have noted in Copernicus, Digges and others. Gilbert had indeed taken a decisive step forward into 'science', but his other foot was still firmly embedded in magic.

When we thus take into account the whole of Gilbert's 'philosophy' we may be less surprised at Bacon's seeming lack of appreciation. More congenial to his approach to natural knowledge, which he believed ought to resemble that of the Presocratic 'physicists', was the system set forth by the South Italian, Bernardino Telesio, under the title *De Rerum Natura*, recalling the great poem by Lucretius, whose influence was then increasing. Bacon praised Telesio for having based his natural philosophy on sensory experience, and censured him for having failed to rid himself of the more speculative flights of Aristotle against which Telesio had claimed to have set his face. More important is Bacon's acceptance of the dual nature of the human soul—one part rational, the other irrational. The study of the former, Bacon himself thought, must 'in the end' be handed over to religion: but the latter (better called the 'sensible soul') is a 'fit subject of enquiry even as regards its substance . . . for the sensible soul—the soul of brutes—must clearly be regarded as a corporeal substance . . . as Bernardinus Telesius and his pupil Augustinus Donius have not altogether unprofitably maintained'. What Telesio had maintained is to be found in the Fifth Book of the *Nature of Things*: here he insisted that not only taste but all sensation (sound excepted) was effectuated by 'touch' (*tractus*). The exception made of sound strikes us as odd, in view of the well-known belief of its being, as Telesio says, 'undoubtedly air in motion' (*commotus nimirum aer*); what he was probably trying to convey was that for instance visual replicas of distant objects actually entered the sensory soul whereas auditory sensations were 'really' impacts of moving air. A similar view is expounded by Lucretius in the theory of *phantasmata*. If this interpretation is correct, Telesio's distinction was perhaps the first impulse (*c.* 1550) during the Renaissance towards the more radical one of primary and secondary qualities, first made explicit by Galileo, Descartes, and Mersenne and subsequently playing a basic role in the empirical philosophy of John Locke. In this distinction touch alone is held to give information of the 'real' nature of external bodies, all other qualities being 'added' by the mind.

Bacon's early *caveat* against the 'mingling' and confounding' of what he called 'philosophy' and 'divinity' here receives a more detailed clarification accompanied by a further warning. While 'philosophy' is restoring the presocratic belief in the materiality of the sensory functions of the human soul it must never be forgotten that there exist 'higher' functions absent from that of beasts: in the latter the sensitive soul is all; in Man it is only the instrument of the rational soul. That this was not a wholly satisfactory demarcation Bacon himself appears to hint. It was of course examined from the complementary point of view, that of the rational soul (*Je pense donc je suis*) about ten years after Bacon's death. The consequential dualism of 'thinking substance' (unique to Man) and 'extended' substance' (the residual empirical finite, including Man's body) unquestionably facilitated the subsequent clarification of the structure and function of the brain and nervous system. But it had other more insiduous consequences that will be referred to in our final assessment. Meanwhile there remains one aspect of the 'great Instauration' that calls for further consideration.

While it is true that the Royal Society was the first to attempt to translate into action the ideal imaginatively created by Bacon in the account of 'Salomon's House' in the *New Atlantis* (p. 161), the basic notion of a continuing group of men engaged in cooperative study of natural phenomena had been realised about the time of Bacon's birth in the 'Academy of Natural Curiosities' that met in Naples for several years under the leadership of Giambattista della

Porta, whose book, *Natural Magic* (1558, enlarged 1589, and many times reprinted and translated), probably formed the basis of their proceedings. The enterprise may well have played a part in extending the idea of an 'academy' from the realm of linguistic and literary studies that were then being pursued in numerous academies in France and Italy to the promotion of natural knowledge. But, as Dr Marie Boas-Hall engagingly puts it, della Porta's 'interest was really that of the party conjurer who deceives the eye by the quickness of the hand'; there is no evidence of the essential aim of making any topic 'a steddy . . . an uninterrupted work' (p. 162). In any case the 'Academy' ran too near the wind and was disbanded after a few years. Torn by religious wars and the tighter discipline of the 'Counter Reformation' the continental countries provided little encouragement for further ventures of this kind. In England on the contrary there were from about 1585 two associated groups of which the patrons were Henry Percy, the so-called 'wizard' Earl of Northumberland and Sir Walter Ralegh. Of the actual proceedings of these groups we know very little; except that during the Earl's confinement in the Tower of London alchemical operations were conducted in a laboratory there and at the same time by Thomas Harriot (p. 109) at the Earl's residence, Sion House, about ten miles west of the City. Harriot was only the most notable of several eminent practitioners in receipt of pensions from one or others of these patrons. The contemporary sobriquet 'wizard' indicates at least a popular belief that 'magic' was afoot; but from what we can learn of Ralegh's character from the style and temper of his *History of the World* written while he was in the Tower it seems unlikely that he at any rate would have been a party to anything but a conservative form of 'natural magic'. From the contents of Percy's great library at Sion we may infer a wide-ranging interest in natural knowledge; from Harriot's especial concern with problems of navigation we may equally infer a severely practical aim; but that such an aim was at that time nowise inconsistent with 'spiritual exercises' of a highly dubious nature we know from the records of John Dee (p. 102). The circumstances in which the Northumberland circle functioned would in any case give rise to rumours and these would have been promoted by the prevailing censorship that prevented the publication of even Harriot's wholly innocuous works on algebra.

However significant may have been the trend of the Northumberland circle in the direction later formulated by Bacon their proceedings lacked one essential characteristic—that of open commerce with the public at large, what has in recent times been called the 'socialisation of science'. This aim was made explicit in the will of Sir Thomas Gresham dated 1575. It is an interesting speculation as to what effect the opening of the 'third university' in the City of

London might have had on the contemporary proceedings in the
Tower and at Sion House. The titles of the endowed chairs—
Rhetoric, Geometry, Astronomy, Music, Divinity, Law, and
Physic (Medicine)—approximating fairly closely the medieval
trivium, quadrivium, and higher faculties (p. 20) reflect a somewhat
conservative vision: 'chymistry' is a notable omission. Conservative
also is the priority given to teaching; but this was inevitable in an
institution whose doors were to be open to anyone who wished to
attend the courses of lectures delivered first in English and repeated
(for the benefit of foreigners) in Latin. By providing the professors
not only with stipends but also resident accommodation Sir Thomas
must surely have had in mind the necessity for sustained study as a
requisite for effective teaching as also the advantage of stimulating
discussion and mutual criticism in a compact society. By ensuring
the location of the College in London the founder, though a loyal
alumnus of Gonville Hall under John Caius (p. 90), tied the activities
of his college to the urgent problems of society in a rapidly growing
mercantile and industrial centre with an opening to the Seven Seas.
Owing to the stamina of his widow, to whom he had left a life-interest
in the property, the inauguration of Gresham College was postponed
till 1597, by which time the famous *Accademia dei Lincei* had been
founded under the patronage of Duke Federigo Cesi in Rome.
attracting such figures as Galileo and John Milton it might have
been the first truly scientific academy; but once again the suspicion
of the ecclesiastical authorities brought about its dissolution.

The institution that approximated most closely the structure of
the Royal Society was the so-called Elizabethan College of Antiquaries: it had a limited membership of 'gentlemen', regular meetings
conducted under formal rules of procedure, and a constitution
setting out its aims. Its composition differed from that of the early
Royal Society in respect of the professional dedication and high
competence of its members. The quality of its membership is reflected
in their proceedings: their aim was to establish 'the facts' about the
relation of existing institutions to those of antiquity; mere 'curiosities'
and snap judgments were discouraged; the emphasis was on documents and the means of their verification. The task of discovering,
transcribing, and reporting on the relevant documents was allocated
to individual members (cf. the 'committees' and their 'curators' in
the Royal Society). This attempt to make of history a 'science'
rather than a literary art, later rejected by Ralegh, but favoured
(at least in theory!) by Bacon was a great step forward in English,
perhaps in all, historiography; as Boyle was a dominant influence
in the early Royal Society, so William Camden was the leading figure
in the 'antiquarian'. Of course their judgments were often wide of
the mark; the generally accepted but partly mythical chronology

of the 'ancient kingdoms' was a stumbling block for another century; but since their emphasis was on the establishment of individual facts rather than on broad conclusions this was of less account. How far this body, whose revival after a lapse of some years King James 'misliked' sufficiently to veto, influenced the movements before the Civil War which culminated in the foundation of the Royal Society it is difficult to say; but that there was much common ground is evidenced by the active and critical interest in antiquities displayed by Newton, Halley, and others.

In the spirit of these Elizabethan antiquaries, but armed with the far wider and more firmly based knowledge to which their new standards of historiography pointed the way, let us attempt to achieve a just perspective of science in a Renaissance society. From whatever standpoint we take up, and in whatever direction we look, we are struck by the acute self-consciousness of Renaissance men. Whether the 'Renaissance' was a historical fact or not there is no doubt that a great many outstanding minds thought that they were living in an age of 'rebirth'. For some historians the reconstruction of what men were saying or thinking during a period under consideration is the best basis for the history of that period; others have been sceptical of this point of view. To the author the history of science, whether 'internal' or 'external' taking account of its social relations, is primarily an aspect of the history of ideas and as such must give at least priority to what was thought and said. If this be granted, it becomes of paramount importance to avoid the facile structuring of Renaissance 'science' in terms of the abstract categories that were in large measure the consequence of its own revaluations of natural knowledge. To ask whether the Renaissance was a historical fact is itself an example of this danger; for to pose the question thus is to risk forgetting that the meaning of the term 'historical fact' was one that the late Renaissance historians and antiquaries were trying to elucidate. Professor F. S. Fussner has persuasively urged that if we find it convenient to speak of a 'Scientific Revolution' during the hundred years following, say, 1580 then we must also recognise that there was at about the same time a 'Historical Revolution'.

Bearing in mind this categorical *caveat* we found that to the oft-debated question as to the relation of the Italian Renaissance to the 'new science' there is no simple answer. Apart from the difficulties raised by the misleading labels 'humanist' and 'technologist', under which rival claimants to the crown of progress have commonly appeared, we have seen that the absence of radically new additions to knowledge does not exclude innovation. The matter is further complicated by the fact that this innovation appears to be founded less on 'scientific' reason than on the myth of an ancient wisdom whose recovery is to be sought by means that had ultimately to be eradicated. The review of the well-charted area of mutual influence of the Italian and 'Northern' centres of scientific activity sprang no surprises except, perhaps, to stress the importance of the monastic complex of Klosterneuburg having a more marked continuity with medieval scholasticism. In a similar way it was found necessary to call in question the traditional judgment of an enlightened 'Platonism' displacing the antique machinery of 'Aristotelianism'. Progress in *natural* science, as distinct from the abstract mathematical relationships of astronomy, took place mainly in the universities of Bologna and Padua—the homes of critical Aristotelianism. The logical basis of seventeenth-century science owes more than has generally been admitted to Giacomo Zabarella, the last of the great Aristotelians at Padua. The refining of this 'method' was exemplified in the proceedings of the Medical Schools.

When we turned to the emerging nation-states we drew attention to the absence of any 'scientists' of even the second rank until at least a century after the awakening of a new spirit in Italy and 'Germany'. This fact was tentatively correlated with the growth towards autocratic centralisation of England, France, and the Iberian Peninsula. Not only was the emergence of science retarded in these circumstances, but the kind of science that did first emerge seems to be related to the structure of government—the kind of science in fact that demands greater resources than are likely to be available in smaller states.

The discovery of America and the 'communication explosion' were seen as emergent technologies: neither owed anything to 'science' but when achieved each had marked consequences for the direction and style of scientific activity. The problem of ruling an immense 'colonial' population under wholly novel conditions of climate, culture, and resources, together with the emergence of new forms of government and experiences of the machinery of credit and exchange, inevitably demanded a change in the political theory underlying the relatively static hierarchical system of medieval Europe. How far the widely influential manifestos of Machiavelli were productive of a more 'scientific' approach or even, as Leopold Olschki claimed, of the 'method' of science itself is still problematic. To say rather that like the urge for geographical expansion it was the product of the *Zeitgeist* is of course no explanation but points to a belief in their common origin. It seems at any rate highly probable that the gradual replacement of 'rational' by 'empirical' theories of sovereignty (the latter backed by a critical and comparative historiography) reveals a world of thought 'much beholden to Machiavel'.

In cosmography the great intellectual achievement of Copernicus was seen as in some degree the product of that rather backward-looking humanism that its subsequent acceptance did so much to discredit. It was Giordano Bruno who, in works published in London before either Galileo or Kepler had put forward any serious views on the subject, shattered the medieval cosmos to fragments and destroyed the distinction between the celestial and 'sublunary' realms. Serious 'science fiction' began with Kepler's *Dream* of a magical voyage to the Moon making possible a rational 'selenography'; this was followed by the more matter-of-fact *Discovery of a World in the Moon* by John Wilkins, chairman of the meeting that called the Royal Society into being. It was this cosmical ruin rather than a growing recourse to 'atomies' as such which must have played the leading part in John Donne's lament that the 'New Philosophy calls all in doubt . . . 'Tis all in pieces, all coherence gone, All just supply and all relation'. Shakespeare was very conscious of it, both in *King Lear* and in the speech of Ulysses in *Troilus and Cressida*, 'Take but degree away . . .'. It was not a long step to Pascal's tortured cry, *Les silences des espaces éternels m'effrayent*.

If it has been found almost impossible to discover when the Renaissance 'began' it is no less difficult, at least in regard to science, to say when it ended. Bacon the 'Great Instaurator' never really grasped the character of the machinery of actual scientific discovery. Those great men who did—Galileo, Gilbert, Kepler, Harvey—were in their several ways linked to the past: their conclusions formed the basis of much of modern science; their methods and outlook were still highly coloured by the mythology of the Renaissance.

Not only in the world of nature had 'all coherence gone' but also in the minds of men: some saw the approach of a 'Brave New World', others only a sickness unto death. As usual in such crises of faith there was in each view a measure of truth. Descartes, like Bacon, saw the possibility of 'making ourselves masters and possessors of Nature'. By assuming a radical dualism between observed nature and Man's self-consciousness he was able to sketch the outline of a possible explanation of the former in purely mechanical terms. By thus submitting the whole of the observable world, including Man's organs of thought and feeling, to the immense power of mathematical formulation he gave to succeeding generations the means of seemingly unending progress in the attainment of the mastery of Man over things; but at the cost of Man's cosmical alienation. It was another Frenchman who saw that 'science without conscience is no other than ruin of the soul'. Four centuries later we can see that it may be the ruin of the world.

BIBLIOGRAPHICAL NOTE ON SOURCES

Even a 'select' bibliography of fully representative sources of this book
would far outrun the available space. In compiling this Note the author
has especially kept in mind the needs of two classes or readers: those
relatively unfamiliar with science, and 'scientists' who may wish to extend
their knowledge of the general historical background of the period.
Since some degree of dogmatism has been unavoidable in the text, those
works on which the discussion of each topic has been largely based are
listed under their appropriate chapter headings. As an aid to reference all
books relevant to more than one chapter have been specified in a General
Section that also contains an annotated guide to a number of works of a
general nature, some of which are not specifically referred to in the text.
The term 'op. cit.' in the Special Section signifies an entry in the General
Section; to avoid ambiguity shortened titles are occasionally used in the
former. Wherever possible the English translation of a work is referred to.
The letters 'PB' (or in the case of the numerous valuable Harper Torch-
books, 'TB') indicates the availability of paperback editions at the time
of writing, but is no guarantee that they were in print in 1971.

Since full documentation is held to be undesirable in this series it is
hoped that the system of reference adopted will reveal at least the proximate sources to which the author has been indebted in writing this book.
To the unseen host who have been his inspiration during over twenty years
of 'addiction' to Renaissance studies he can do no more than express a
lasting gratitude in general terms; it is hoped that at least some of the
light they have shed will be reflected here and there in his pages.

GENERAL SECTION

The basis of all later discussion is of course Jacob Burckhardt, *The
Civilisation of the Renaissance of Italy*, of which several editions of the
translation by S. G. C. Middlemore are available (e.g. Mentor PB, New
York 1961, and Phaidon PB, London). The most comprehensive study

of the 'Problem of the Renaissance' is Wallace K. Ferguson, *The Renaissance in Historical Thought* (Cambridge, Mass. 1948); a later brief statement by Ferguson occurs in *Facets of the Renaissance* (Harper TB, 1963) which also contains a vigorous essay relevant to the question by Garrett Mattingly. A set of essays on changing interpretations of various aspects of the Renaissance is comprised in *The Renaissance*, Ed. Tinsley Helton (PB, Madison, Wis. 1964).

For the general historical background the following are recommended: for the fifteenth century, M. Gilmore, *The Age of Humanism* (New York 1952) and Margaret Aston, *The Fifteenth Century* (Thames and Hudson PB, London 1968); for the sixteenth century, H. G. Koenigsberger and G. L. Moss, *Europe in the Sixteenth Century* (London 1968).

The best introduction to the scientific background is M. Boas Hall, *The Scientific Renaissance 1450–1630* (London 1962). The greater part of A. R. Hall, *The Scientific Revolution 1500–1800* (London 1954) goes beyond our period, but the earlier chapters give a very good idea of the state of scientific knowledge at the beginning of the sixteenth century. Somewhat more detailed than either of these is W. P. D. Wightman, *Science and the Renaissance* (Aberdeen University Studies no. 143, Edinburgh 1962), vols. I (narrative) and II (a complete annotated bibliography of the representative collection of contemporary—mainly sixteenth-century—books in the University of Aberdeen); vol. 1 contains a list of secondary works.

The bibliographical aspect of Renaissance science is sympathetically treated by G. Sarton in *Six Wings* (Indiana 1957, London 1958). The more philosophical attitudes are well represented in *Theories of Scientific Method*, Ed. E. H. Madden (University of Washington PB, Seattle and London 1966); the first four chapters are relevant, of which ch. II on the *Hypotheses of the Renaissance Astronomers* is particularly valuable.

Changing attitudes to nature and human experience may be studied in two candid autobiographical writers: the earlier credulous genius, Girolamo Cardano, *The Book of My Life by Jerome Cardan* (Trans. Jean Stoner, London 1931) and the *Essays* of the sceptical Michel de Montaigne (Trans. with an introduction by J. M. Cohen, Penguin Classic, 1958).

A. V. Martin, *Sociology of the Renaissance* (Trans. from German ed. of 1932, London 1944), is important for the questions raised rather than for the answers attempted in such small compass.

Several of the works of Eugenio Garin have been heavily drawn upon, but only one has been translated into English, *Italian Humanism* (Oxford 1966), more wide ranging than the title might imply.

Muir's Historical Atlas—Medieval and Modern (10th ed., London 1964, based on the redesigned ed. of 1962) is very useful as a guide to changes in territorial boundaries during this difficult period.

Finally, the extensive bibliography of Renaissance studies contained in F. Chabod, *Machiavelli in the Renaissance* (Harper TB, 1965), though compiled in 1957 and dealing mainly with the Italian Renaissance, is nevertheless of great value. The article on *The Concept of the Renaissance* in the same volume is also most valuable.

The following list of books later referred to under more than one chapter heading is appended for convenience in reference:

Alberti, L. B., On *Painting*, Trans. and Ed. J. R. Spencer, London 1956

Caspar, Max, *Kepler*, Trans. and Ed. D. Hellman, Collier PB, New York 1952

Crombie, A. C., *Robert Grosseteste and the Origins of Experimental Science*, Oxford 1953

Debus, A., *The English Paracelsians*, London 1965

Documentary History of Art, Ed. E. Holt, Princeton 1947; Doubleday Anchor PB, 1957

Dresden, S., *Humanism in the Renaissance,* Trans. M. King, Weidenfeld & Nicolson PB, London 1969

Elliott, J. H., *The New World and the Old*, Cambridge PB, 1970

Garin, E., *Italian Humanism*, Oxford 1966

Hall, M. Boas, *Nature and Nature's Laws*, Harper TB, New York 1970

Ivins, W. M., Jr., *Art and Geometry*, Harvard 1946; Dover PB, 1964

Keller, A. G., *Theatre of Machines*, Chapman & Hall, London 1964

Machiavelli, N., *The Prince*, Trans. G. Bull, Penguin reprint 1966

Mattingly, G., *Renaissance Diplomacy*, London 1955; Peregrine PB, Penguin, 1965

Morison, S. E., *Columbus Admiral of the Ocean Sea*, London 1942

Nicholas of Cues (Nicolaus Cusanus), *Of Learned Ignorance*, Trans. and Ed. G. Heron, London 1954

Ong, W. J., *Ramus, Method and the Decay of Dialogue*, Cambridge, Mass. 1958

Pagel, W., *Paracelsus* (in English), Basel 1958

Ridolfi, R., *Life of Nicolò Machiavelli*, Trans. C. Grayson from the Italian, Rome 1954; London 1963

Rossi, P., *From Magic to Science*, Trans. F. Rabinovitch from the Italian, Bari 1957; London 1968

Rossi, P., *Philosophy, Technology and the Arts in the Early Modern Era*, Harper TB, 1970

Taylor, E. G. R., *Tudor Geography 1485–1583*, London 1930

Taylor, E. G. R., *The Mathematical Practitioners of Tudor and Stuart England*, Cambridge 1954

SPECIAL SECTION

Chapter 1: The Renaissance and science in Italy

Burckhardt, op. cit.

Baron, H., *The Crisis of the Early Italian Renaissance*—an abridged one-volume edition of an earlier larger work. PB, Princeton, 1966. For a roughly parallel growth of urban communities see H. Pirenne, *Early Democracies in the Low Countries*, Harper TB, 1963. The conclusions of the original work are no longer wholly valid, as indicated in the Introduction.

Alberti, L. B., op. cit. Joan Gadol's *Leon Battista Alberti, Universal Man of the Early Renaissance*, Chicago 1969, was not available until after this book was almost completed. See Professor C. Grayson's review in *Renaissance Quarterly* 24 (1, 1971), 51.

Aston, M., op. cit.

Mattingly, G., op. cit., especially Part II.

Various aspects of humanism and the philosophical background of the Italian Renaissance are dealt with by P. O. Kristeller in *Renaissance Thought* (1961) and *Renaissance Thought II* (1965), Harper TBs.

Garin, E., op. cit.

Chapter 2: *Science and the Northern Renaissance*

This chapter is based mainly on Willy Andreas, *Deutschland vor der Reformation*, 5th ed. Stuttgart 1948 (1st ed. 1932) and O. Benesch, *The Art of the Renaissance in Northern Europe*, Cambridge, Mass. 1947.

For Nicholas of Cues see his own great work *Of Learned Ignorance* (Trans. G. Heron, London 1954) and, for his great influence on Renaissance thought, E. Cassirer, *The Individual and the Cosmos* (Oxford 1963, Trans. from the original *Individuum und Kosmos in der Philosophie der Renaissance*, Leipzig 1927), one of the great pioneering works of reinterpretation of the Renaissance spirit and containing other articles and extracts.

Durand, D., *The Vienna-Klosterneuburg Map Corpus of the Fifteenth Century*, Leiden 1952, provides a most illuminating picture of the activity in the monastic complex of Southern Germany during the early decades of the fifteenth century. It is fully documented and supported by detailed study of astronomical and cartographical MSS.

Waetzoldt, W., *Dürer and his Times*, Trans. R. H. Boothroyd, Phaidon, London 1950.

Rossi, P., *Philosophy etc.*, op. cit. A not entirely satisfactory translation of several valuable earlier articles collected in an Italian edition Milan 1962.

Chapter 3: *'Men can do all things if they will'*

Alberti, op. cit.

The *Documentary History of Art*, op. cit., provides translations of numerous extracts illustrating the work of the Italian 'artist-engineers'. Kenneth Clark, *Piero della Francesca* (2nd revised ed., London 1969), provides a detailed study of one of the most important of them. For perspective and 'mathematical-harmonic' architecture see Ivins, op. cit., and R. Wittkower, *Architectural Principles in the Age of Humanism* (Tiranti, London 1962, finely illustrated PB) respectively.

Rotondi, R., *The Ducal Palace of Urbino* (Tiranti, London 1969), is a somewhat abbreviated translation of the definitive Italian work (1958) by the Curator of this wonderful creation sponsored by Federigo da Montefeltro—very fully illustrated.

The earlier chapters of A. O. Lovejoy, *The Great Chain of Being* (Harper TB, 1960; original ed. New Haven, Conn. 1936) provide a good introduction to Neoplatonic cosmology recognisable throughout the Middle Ages and becoming a dominant influence in the Renaissance.

Chapter 4: 'The Gutenberg Galaxy'

The early development of printing is fully described in G. H. Putnam, *Books and their Makers during the Middle Ages*, two vols. of which the first covers 476–1600 (New York 1898).

An excellent review of the printed 'scientific' literature of the Renaissance is G. Sarton, *The Appreciation of Ancient and Medieval Science during the Renaissance 1450–1600* (University of Pennsylvania 1955; Perpetua PB, New York 1961).

Though its thesis may have to be drastically modified and despite the fact that its exposition is often misleading as to details Marshall McLuhan's *The Gutenberg Galaxy* (London and Toronto 1962; Routledge PB, 1967) is in a sense an epoch-making work and should be regarded as essential reading.

Ong, W. J., op. cit., has sections highly relevant to this chapter.

Chapter 5: Science and the New World

Morrison, S. E., op. cit., differs from the two-volume definitive edition published at the same time only in respect of the documentation, some navigational tables, and the abbreviation of two chapters. For some critical remarks on the various accounts, by different authors, of the relation of Columbus with Toscanelli, see Wightman, op. cit., vol. I, pp. 129f.

This chapter owes a great deal to the collection of excerpts from contemporary sources published and edited with valuable introductory sketches by J. H. Parry in *The European Reconnaissance* (London 1968).

Elliott, J. H., op. cit., appeared only after this chapter had been written.

The general context of discovery is discussed by H. V. Livermore and J. H. Parry (both ch. XV) and by E. E. Rich (ch. XVI) in the *New Cambridge Modern History*, vol. I, 1957.

For the technical aspects of navigation see E. G. R. Taylor, *Tudor Geography 1485–1583*, op. cit., and *The Haven-finding Art* (London 1956). Also, Waters, D. W., *The Art of Navigation in England in Elizabethan and Early Stuart Times*, London and Yale U.P., 1958

Chapter 6: Science and political theory

The point of departure for this chapter was L. Olschki's challenging *Machiavelli the Scientist* (Berkeley, Calif. 1945). For Machiavelli himself see R. Ridolfi, op. cit., and for a critical appraisal of his relation to Renaissance (mainly political) thought the articles by F. Chabod referred to in the General Bibliography. A translation of

The Prince is available in Penguin Books. Garin, op. cit., has numerous incidental remarks on Machiavelli.

The question of the 'Demarcation Lines' is dealt with by Morison, op. cit., also by Livermore (loc. cit. in *New Cambridge Modern History*).

Mattingly, G., op. cit., provides an admirable background for the whole discussion.

The article by Felix Gilbert, 'Bernardo Rucellai and the Orti Oricellari' (*Journal of the Warburg and Courtauld Institutes XII*, 1949) shows that an empirical approach to political theory was being discussed in Florence about ten years before *The Prince*. W. H. Greenleaf, *Order, Empiricism and Politics* (London 1964), discusses the gradual replacement, at the end of the Renaissance, of 'rational' by empirical theories of sovereignty, mainly in respect of England.

The later pages of this chapter bear evidence of the author's debt to Elliott, op. cit.

Chapter 7: Innovation and method in Medicine

The basic work on 'method' is J. H. Randall, Jr, *The School of Padua and the Emergence of Modern Science* (Padua 1961). For some amplification and a measure of criticism for the highly important thesis developed by Randall see Wightman, '*Quid sit Methodus?* Method in sixteenth century teaching and discovery', *Journal of the History of Medicine XIX* (1964), 360–76.

The logical aspects are treated by Crombie, op. cit.

For Paracelsus, see Pagel, op. cit.

The entry BENIVIENI in the *Dictionary of Scientific Biography* (New York 1970) puts his work in a clearer light than has usually been the case heretofore.

Chapter 8: Science in the emergent nation-states

For the general historical background to the discussion in this chapter Mattingly, op. cit., is as usual basic.

For France, see D. Stone, Jr., *France in the Sixteenth Century—a Medieval Society transformed* (Prentice-Hall PB, Englewood Cliffs, N.J. 1969); the (mainly literary) culture of the century is reviewed against the background of political history. This may be compared with the appropriate chapters in S. Dresden, op. cit. Sir John Neal's *The Age of Catherine de Medici* is indispensable (1943; Jonathan Cape PB, 1963).

For England C. R. Elton *England under the Tudors* (London 1955) has been found most helpful.

The intellectual climate in England during the sixteenth century is still a subject of controversy. J. H. Hexter *Reappraisals in History* (1961; Harper TB, 1963) makes two critical contributions, *The Education of the Aristocracy in the Renaissance* and *The Myth of the Middle Class in England*. The impact of the Reformation is presented from different points of view in *The Reformation Crisis*, Ed. J. Hurstfield (Arnold PB, London 1965; rep. 1968).

The growth of science in England from about 1580 was clarified in the important book *Astronomical Thought in Renaissance England* by F. R. Johnson (Baltimore 1937). E. G. R. Taylor, op. cit., was one of the first to stress the importance of individual centres such as John Dee. The earlier chapters of C. E. Raven *English Naturalists from Neckam to Ray* (Cambridge 1947) tell of the relations of English naturalists with such continental figures as Conrad Gesner, Luca Ghini, etc. *Oxford and Cambridge in Transition 1558–1642* by M. H. Curtis (Oxford 1959) is still the most reliable basis for assessing the significance of the universities. A recent work by H. Kearney, *Scholars and Gentlemen* (London 1970), contains new information and examines more closely the impact of Ramism; but the exposition seems to have been somewhat biassed by an attempt to show the dominance of the court 'Establishment'.

Ong, W. J., op. cit., is very useful in respect of the educational issues raised in this chapter. See also Joan Simon, *Education and Society in Tudor England* (Cambridge 1966).

Chapter 9: The Copernican revolution

The somewhat revised standpoint adopted in this chapter owes much to the excellent (though somewhat technical) article by O. Neugebauer, 'On the planetary theory of Copernicus' in *Vistas in Astronomy X* (1968), 89f. Good general accounts are contained in T. S. Kuhn, *The Copernican Revolution* (Cambridge, Mass. 1957) and A. Armitage, *The World of Copernicus* (New York 1963 and later PB).

The necessarily brief and highly selective account of Giordano Bruno is based on P.-H. Michel, *Le Cosmologie de Giordano Bruno* (Paris 1962). For the life and intellectual background of Bruno, F. Yates' *Giordano Bruno and the Hermetic Tradition* (London 1964) is very wide-ranging and likely to remain a definitive work.

For some useful extracts see M. Boas Hall, op. cit.

Chapter 10: Accent on quantity

The following works are of special significance for this chapter:

Nef, J. U., *Cultural Foundations of Industrial Civilization*, Cambridge 1958; Harper TB, 1960.

Mattingly, G., *The Defeat of the Spanish Armada*, London 1959 and several reprints.

Donald, M. B., *Elizabethan Monopolies*, Edinburgh 1961.

Ore, O., *Cardano the Gambling Scholar*, Princeton 1953.

Dreyer, J. L. E., *Life of Tycho Brahe—a picture of scientific life and work in the sixteenth century*, Edinburgh 1890. Parts of this book provide a vivid idea of the new standards of precision and organisation of scientific work.

The translation of Copernicus's tract on money is by J. Taylor in *Journal of the History of Ideas* 16 (1955), 540f.

The article by A. Koyré, 'La dynamique de Nicolò Tartaglia', is reprinted in Koyré's book *Etudes d'Histoire de la Pensée Scientifique* (Paris 1966), a valuable collection by René Taton after Koyré's death.

Chapter 11: From magic to science

For Bacon see the recommendations under Chapter 12. *History of Magic and Experimental Science* by Lynn Thorndike was a pioneering work in this field and contains a great deal of valuable information. Vols. IV and V–VI are on the fifteenth and sixteenth centuries respectively.

The *Oration on the Dignity of Man* by Giovanni Pico della Mirandola is translated by E. L. Forbes and introduced by P. O. Kristeller in *The Renaissance Philosophy of Man*, Ed. E. Cassirer, P. O. Kristeller and J. H. Randall, Jr. (Chicago 1948; PB 7th imp., 1961). Here also is to be found Pietro Pomponazzi's *On the Immortality of the Soul,* translated and edited by Kristeller, Randall and Hay. For a discussion of Pomponazzi see Cassirer, *Individual and Cosmos*. . . .

The English translation of the *Corpus Hermeticum* is by W. Scott under the title *Hermetica*, 4 vols. (Oxford 1924–6). The critical notes of A. S. Ferguson in vol. 4 are very valuable but the (French) translation of A. D. Nock and A.-J. Festugière is said to be more satisfactory than Scott's.

The charting of the various views on magic during the Renaissance has nowhere been better carried out than by D. P. Walker in *Spiritual and Demonic Magic from Ficino to Campanella*, Studies of the Warburg Institute no. 22 (London 1958).

For Kepler's analogies see *A New Year's Gift—The Six-Cornered Snow-flake*. Text and trans. from the Latin (Frankfurt 1611) by C. Hardie with additional essays by L. L. Whyte and B. J. Mason (Oxford 1956). Also Caspar, op. cit.

An adequate study of Cornelius Agrippa is now available in C. G. Nauert, *Agrippa and the Crisis of Renaissance Thought* (Urbana, Ill. 1965). Though somewhat repetitive this book effectively suggests the flavour of the later Renaissance.

Benedetto Varchi's *Questione sull' Alchimia* (modern ed., Florence 1827) is of great interest in relation to the climate of opinion in mid-*cinquecento* Italy.

Chapter 12: 'The Great Instauration'

The definitive source for Francis Bacon is *The Works of Francis Bacon*, Ed. with still perceptive introductions by R. L. Ellis, J. Spedding, D. D. Heath (London 1887–92). The most comprehensive book on Bacon's influence on science is Paolo Rossi, *Francesco Bacone Della Magia alla Scienza* (Bari 1957), Trans. S. Rabinovitch, *Francis Bacon —From Magic to Science* (London 1968). B. Farrington's *The*

Philosophy of Francis Bacon—Essays on Development 1603–9 (London 1964) is important in respect of the three translations of youthful, more spontaneous views.

Catherine D. Bowen's *Francis Bacon—The Temper of a Man* (London 1963) provides a very sympathetic portrait of the man himself.

The best introduction to W. Gilbert's work is D. H. D. Roller, *The De Magnete of William Gilbert* (Amsterdam 1959). The only translation available is by P. F. Mottelay (London 1893 and recent PB) but it is not altogether satisfactory.

The account of the Kepler–Fludd–Mersenne controversy and its ramifications is based on D. P. Walker (op. cit., under Chapter 11) and his article on 'Kepler's planetary music' in *Journal of the Warburg and Courtauld Institutes XXX* (1967). In the same volume there is also P. J. Ammann's *Musical theory and philosophy of Robert Fludd*. A. C. Crombie's very important 'Mathematics, Music and medical science' (*Organon 6*, 1969) will be incorporated in a forthcoming book (private communication).

For the other topics discussed in this chapter the following are recommended:

Outhwaite, R. B., *Inflation in Tudor and Early Stuart England*. This is a recent 'case-history' relevant to the general European crisis (Papermac published for the Economics History Society, 1969).

Fussner, F. S., *The Historical Revolution—English Historical Writing and Thought 1580–1640* (London 1962). Space did not permit of any development of the important thesis presented in this work. Similar remarks may be made in respect of W. H. Greenleaf, *Order, Empiricism and Politics—Two traditions of English Political Thought 1500–1700* (London 1964).

Thomas Sprat's *History of the Royal Society of London for the Improving of Natural Knowledge* (London 1667 and later eds.) is available in a reprint edited with critical apparatus by J. I. Cope and H. W. Jones (St Louis, Missouri 1958).

There is not much in English on Bernardino Telesio, but P. O. Kristeller summarises his system, also that of the contemporary Francesco Patrizzi, in *Eight Philosophers of the Italian Renaissance* (Stanford, Calif. 1964). Good notes and bibliography.

C. Hill's *Intellectual Origins of the English Revolution* (Oxford 1965) contains a mine of information relating to our period, but the thesis it is used to support does not command general assent.

G. Della Porta's famous work, trans. in 1669, was reprinted as *Natural Magic in 20 Books*.

INDEX OF NAMES

This index consists mainly of persons and places: under the latter will be found all important references to the universities. Only points of special interest have been included under (e.g.) 'France', etc. The dates of emperors, kings and popes are *regnant* years; of Italian 'Princes' and of Catherine de' Medici life-spans (Catherine was queen-consort and, after 1560, regent of France). Where no basis of exact dating is available the Sarton 'century method' has been used, e.g., 'II–1 B.C.' means 'first half of the second century B.C.'

GENERAL INDEX

H1